T3-BLC-391

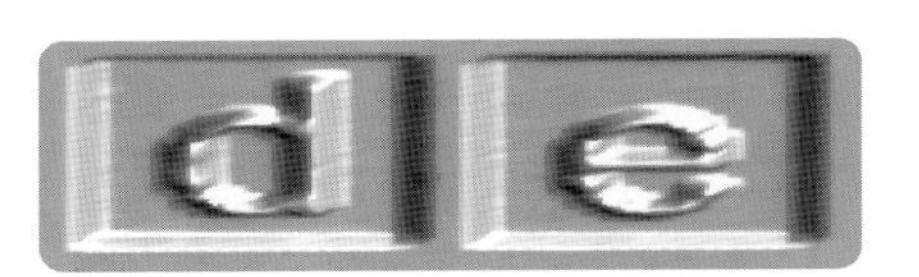

sign

projektarbeiten des fh-studienganges industrial design-graz 1996/2001

projects of the fh-degree course industrial design-graz 1996/2001

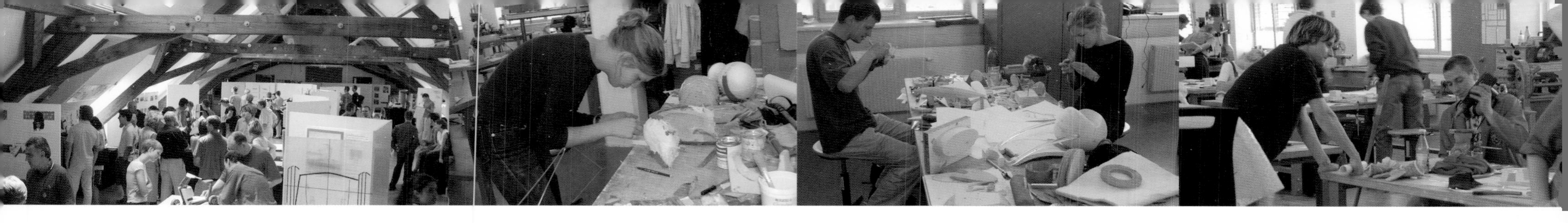

--- 66 studierende --- 7 nationen --- 100-140 bewerbungen für 18 studienplätze ---

--- 66 students --- 7 nationalities --- 100-140 applicants for 18 places --- 8 semesters ---

--- 8 semester --- praxissemester integriert --- dipl.–ing. für industrial design (fh) ---

--- integrated semester of practical training --- master of science (fh) in industrial design ---

impressum/*imprint:*

copyright	studiengang industrial design fh joanneum, graz alle rechte vorbehalten/*all rights reserved*
herausgeber *editors*	gerhard heufler/gerald kiska fh joanneum gmbh, graz
konzeption *concept*	gerhard heufler/gerald kiska
redaktion *editor in chief*	gerhard heufler
übersetzungen *translations*	eva dringel-techt und design studenten *eva dringel-techt and authors*
gestaltung *design*	martin prettenthaler www.urban-filter.com
bildrechte *photos and illustrations by permission of*	fh-studiengang industrial design audi design (seite 66/67)/wallpaper (seite 139)
buchbinder *binding*	buchbinderei schumacher ag, schmitten
adresse *address*	studiengang industrial design/fh joanneum alte poststraße 149, a-8020 graz/austria tel: +43-316-5453-8100/fax: +43-316-5453-8101 industrial.design@fh-joanneum.at
web	http://industrial-design.fh-joanneum.at

© 2002 by verlag niggli ag, sulgen | zürich
isbn 3-7212-0446-8

unterstützt und empfohlen von:
sponsored and recommended by:
» die österreichische designstiftung «
» wirtschaftsressort der steiermärkischen landesregierung «
» wissenschaftsressort der steiermärkischen landesregierung «

design products

projektarbeiten der fh graz 1996 – 2001

herausgeber / *editors*
gerhard heufler / gerald kiska / fh joanneum

verlag / *publisher*
niggli ag

vorwort / *preface:*

›› der nährboden für innovative designer ‹‹

mit den fachhochschul-studiengängen "industrial design" und "informationsdesign" wurde an der fh joanneum in graz bewusst ein ausbildungsschwerpunkt für design angesiedelt. das ziel war und ist es, die professionalisierung der heimischen designszene mit kreativen kräften zu unterstützen, die nicht nur gestalterisch hervorragend ausgebildet sind, sondern auch das notwendige technische und wirtschaftliche verständnis aufweisen. das innovative österreichische fh-studium bietet dafür die geeigneten voraussetzungen. zu seinen eckpfeilern gehören eine regelstudiendauer von 8 semestern, eine begrenzte zahl von studienplätzen, die eine intensive zusammenarbeit von lehrenden und studierenden fördert, sowie die strikte praxisorientierung. dazu gesellt sich das bekenntnis zur qualität als oberstes credo der fh joanneum.
als anno 1999 die ersten absolventinnen und absolventen von "industrial design" ihre beeindruckenden diplomarbeiten im rahmen einer "degree show" der öffentlichkeit präsentierten, zeigte sich, dass sich die in diese ausbildung gesetzten hoffnungen mehr als erfüllt haben. wir sind zuversichtlich, dass die betrachter dieses buches angesichts der kreativen leistungen unserer studierenden diese anschauung teilen.

›› a breeding ground for innovative designers ‹‹

with the university of applied science degree programs "industrial design" and "information design", a conscientious emphasis on design has been established at the fh joanneum in graz. the objective was and still is to support the professionalization of the domestic design scene through training creative people by providing them not only with outstanding skills in design but also the necessary technical and business knowledge. this innovative austrian university of applied science program offers the appropriate conditions for this. among its corner stones are a regular duration of study of 8 semesters, a limited number of places on the course – demanding an intensive collaboration between students and teaching staff, and a strict practical orientation. added to this is the acknowledgement of quality as the supreme principle of the fh joanneum.
as the first graduates of industrial design presented their impressive diploma theses to the public in the context of a degree show in 1999, it became obvious that the hopes placed in this training course were more than fulfilled. we are confident that the reader of this book will share this view with respect to the creative achievements of our students.

mag. martin pöllinger
geschäftsführung der fh joanneum
management of fh joanneum

mag. dr. peter reininghaus

›› form follows function ‹‹
helmut lang kreiert in new york die weltweiten modetrends, erwin himmel designt in barcelona grosse automarken. beide sind steirer. purer zufall?...
design ist mehr als produkt-facelifting - design ist eine dienstleistung im spannungsfeld zwischen funktion und form. ohne ausgefeiltes design lässt sich heute kein produkt mehr verkaufen. design ist zum innovations- und marketinginstrument nummer 1 geworden. durch die gezielte vernetzung von kreativen und unternehmern soll das kreative potenzial im land in zukunft noch besser zur geltung kommen. die nutzung der kreativen kräfte von design zur darstellung von land, region, unternehmen, produkten und leistungen ist eine einmalige chance für die steiermark, um im internationalen wettbewerb mit dem faktor unterscheidbarkeit einen vorsprung zu erreichen.
das vorliegende buch "design products" – fh-projektarbeiten 1996-2001 stellt eine weitere verankerung des design-themas dar und soll sowohl als impulsgeber für einen weiteren mutigen weg als auch als nachschlagewerk und ideenlieferant verstanden werden, um weitere design-initiativen zu entwickeln und neue identitäten zu schaffen.

›› form follows function ‹‹
in new york, helmut lang is establishing worldwide fashion trends; in barcelona, erwin himmel is designing cars for major automobile makers. both are styrians. is it just a coincidence?...
design is more than just product face-lifting - design is a service in the conflict-laden field between function and form. today, it is simply impossible to sell a product without having a polished design. design has become the number one instrument for innovation and marketing. with the development of a network of creative people and business people, the creative potential in the region should come into its own. the utilization of design's creative and planning strengths to present the character of the country, the region, the businesses, and the products and services is a unique opportunity for styria to attain an advantage in international competition by being distinctive.
this book, "design products" - fh projects 1996-2001, is yet another embodiment of the theme of design. besides being a reference work, it will hopefully generate new impulses for people to persevere on this courageous path and supply ideas so that new initiatives can be developed and new identities created.

dipl.-ing. herbert paierl
landesrat für wirtschaft, finanzen und telekommunikation
landesrat of the department of trade, commerce and telecommunications

›› design hat nicht nur in kunst und kultur bedeutung ‹‹
...sondern ist auch ein wichtiger wirtschaftsfaktor, der gerade in globalen märkten immer mehr an bedeutung gewinnt. um diesem anspruch gerecht zu werden braucht es gut ausgebildete designer, die die sprache der unternehmer sprechen und den unterschied zwischen idee und produkt begreifen. bemerkenswerte beispiele, die diesen anspruch untermauern, finden sie in diesem vorliegenden "design products".
um den an der fh joanneum ausgebildeten designern ein erfolgreiches betätigungsfeld zu geben und um design als einen, für die identität und den wirtschaftlichen erfolg eines landes, bedeutenden faktor zu etablieren, hat nun die österreichische wirtschaft gemeinsam mit wichtigen institutionen ein klares zeichen gesetzt: die gründung der österreichischen designstiftung. die österreichische designstiftung soll helfen, den wert von design zu kommunizieren, sie soll die plattform für österreichisches design werden und sich damit zum ansprechpartner sowohl für unternehmen und designer als auch für ausbildungsstätten und die öffentliche hand entwickeln. dabei werden publikationen wie die vorliegende einen grossen wert erweisen: als beweis aufgeweckter kreativität, als stimulus für die innovativen unternehmer dieses landes oder als erbauliche lektüre für alle design interessierte.

›› design has not only found its place in the arts and in cultural life ‹‹ ...but constitutes a considerable economic factor of increasing importance throughout the global market. in order to fully meet this challenge, it takes well-trained and skilled designers, speaking the language of the companies and who fully comprehend the difference between an idea an the final product. in this "design products", please find remarkable examples that underline these high standards.
in order to allow the designers, following their training at the fh joanneum, to apply their skills, and in order to establish design as an important factor for the identity and the economic success of a country, the austrian economy, in alliance with leading institutions, has set an example in creating: the austrian design foundation. the austrian design foundation is designed to help and communicate the value of design. it will be the leading platform for austrian design and therefore emerge to be the contact for both companies and designers, as well as training centers and the public sector. publications such as this one, will be of vital importance. they constitute a proof of intelligent creativity, a stimulus for the country's leading brains or serve as interesting reading material for all those interested in design.

dipl.-ing. stefan pierer
vorstandsvorsitzender der oesterreichischen designstiftung
chairman of the board of the austrian designfoundation

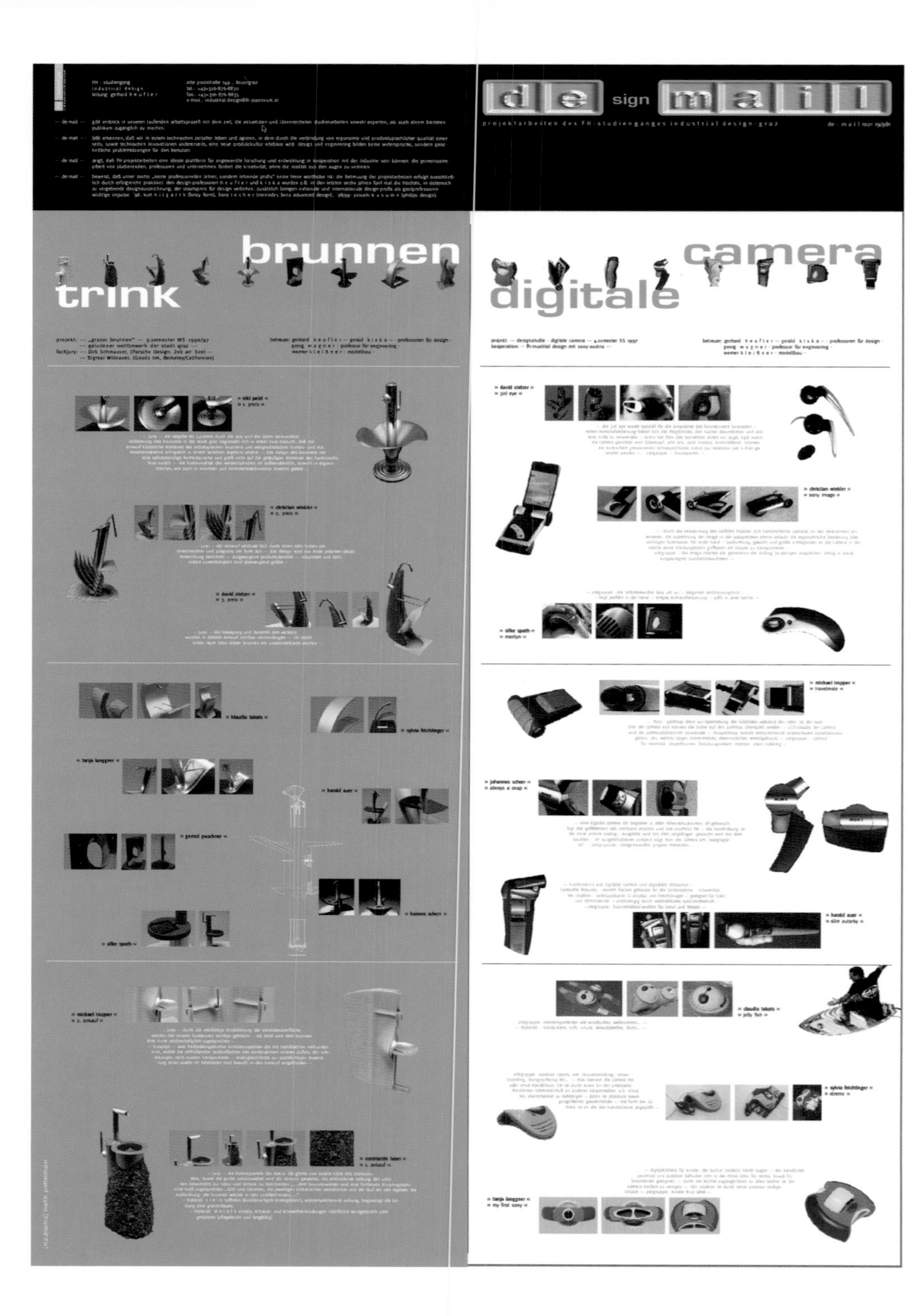

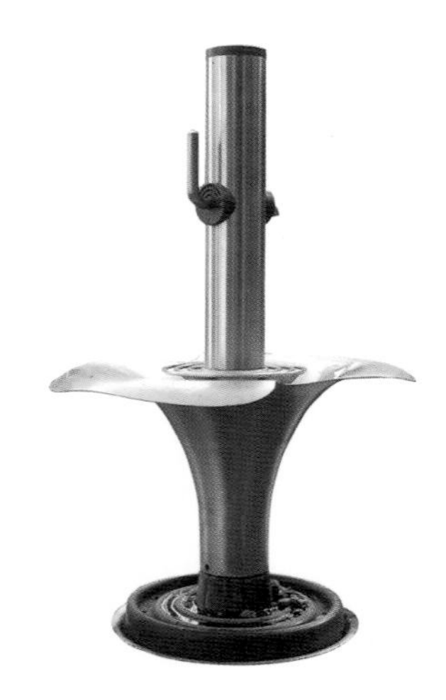

realisiertes siegerprojekt: "grazer brunnen"
von niki pelzl
winning project already realized: "graz fountains"
by niki pelzl

öffentlichkeitsarbeit: die ersten projektarbeiten als poster (59/84 cm)
public relations: the first projects as posters (59/84 cm)

1996 – also bereits ein jahr nach seiner gründung – startete der fh-studiengang industrial design in graz die erste projektarbeit: ein geladener wettbewerb der stadt graz zum thema "trinkwasserbrunnen". es war ein buchstäblich mutiger sprung ins wasser, aber auch ein erfolgreicher: das unter dem juryvorsitz dirk schmauser's von porsche design gekürte siegerprojekt wurde inzwischen realisiert und bereichert auf mehreren grazer plätzen das stadtbild.
seither bilden – nach zwei grundlagensemestern – die vier einsemestrigen projektarbeiten das rückgrat der ausbildung, ergänzt durch ein umfassendes fächerbündel und ein integriertes praxissemester. das vielschichtige lehrangebot reicht von der psychologie bis zum marketing, von der ergonomie bis zur konstruktionslehre und von der mikroelektronik bis zur ökologie. die betreuung erfolgt interdisziplinär, unter einbeziehung von gastprofessoren. unser motto: «keine professionellen lehrer, sondern lehrende profis».
jede projektarbeit publizieren wir inzwischen in form eines posters mit dem titel "de-sign-mail".
die folgenden grundsätze haben unser konzept geprägt:

- "de-sign-mail" gibt einblick in unseren laufenden arbeitsprozess mit dem ziel, aktuelle und ideenreiche studienarbeiten sowohl experten als auch einem breiteren publikum zugänglich zu machen.
- "de-sign-mail" lässt erkennen, dass wir in einem technischen zeitalter leben und agieren, in dem durch die verbindung von ergonomie und produktsprachlicher qualität einerseits sowie technischen innovationen andererseits eine neue produktkultur erlebbar wird: design und engineering bilden keine widersprüche, sondern ganzheitliche problemlösungen für den benutzer.
- "de-sign-mail" zeigt, dass fh-projektarbeiten eine geeignete plattform für angewandte forschung und entwicklung in kooperation mit der industrie darstellen. die gemeinsame arbeit von studierenden, professoren und unternehmen fördert die kreativität, ohne die realität aus den augen zu verlieren.

die regelmässig erscheinenden poster haben innerhalb weniger jahre zu einem hohen bekanntheitsgrad und auch internationaler anerkennung des grazer studienganges geführt. "de-sign-mail" findet man inzwischen an den pinnwänden renommierter designbüros von münchen bis new york.
nach erst vierjährigem bestehen wurde unser studiengang – als erste designhochschule österreichs – vom "bureau of european designers' associations" beda als "high standing educational institution" klassifiziert.

1996-2001 – fünf Jahre projektarbeit am studiengang industrial design in graz sind deshalb ein anlass, bilanz in form dieses buches mit den gesammelten "de-sign-mails" zu ziehen.

gerhard heufler/gerald kiska

1996 – *one year after being founded, the fh course of study for industrial design in graz started its first project – a competition by invitation of the city of graz on the subject of "drinking fountains". it was literally a courageous plunge, but it was also a successful one. with a jury chaired by* dirk schmauser *porsche design, the winning project has already been realized, and there are drinking fountains on several squares in graz, which enhance the cityscape.*
after two semesters filled with basics, four projects of one semester each are the backbone of the course of study, which is supplemented by an extensive number of classes and an integrated semester of practical training. there is a multilayered selection of classes from psychology to marketing, from ergonomics to constructional training and from microelectronics to ecology. student support is interdisciplinary and includes guest professors. our motto: «... no professional teachers but teaching professionals»

we publish each project as a poster with the title "de-sign-mail".
the following principles characterize our concept:

- *"de-sign-mail" gives insights into our work process with the goal of making current, creative student projects available both to experts and a broader public.*
- *"de-sign-mail" makes it clear that we are living and operating in a time period characterized by technology, in which the connection of ergonomics and product quality on the one hand, as well as technical innovations on the other make it possible to experience a new product culture. design and engineering are no longer contradictions, but holistic solutions to user problems.*
- *"de-sign-mail"shows that fh projects are an appropriate platform for applied research and development in collaboration with industry. the joint work of students, professors, and companies promotes creativity without losing sight of reality.*

the posters, which appear on a regular basis, have made the graz course of study well known and have also brought international recognition.
"de-sign-mail" can be found on bulletin boards in renowned design company offices from munich to new york.
only four years after it was founded, our design institution was the first in austria to be classified as a "high standing educational institution" by the "bureau of european designers' associations" (beda).

1996-2001 – *five years worth of projects for the course of study industrial design in graz are a special occasion and a reason to take stock in the form of this book with a collection of all "de-sign-mail" posters.*

gerhard heufler/gerald kiska

anmerkungen zum designprozess:
comments on the design phase:

designstudie: "digitale camera für kinder"
von tanja langgner
design project: "digital camera for children"
by tanja langgner

collage: wertewelt der zielgruppe "6- bis 12-jährige kinder"
collage: values of target group "children aged 6-12"

in der berufsrealität ist der designprozess ein äusserst komplexer, ganzheitlicher problemlösungsprozess. er spielt sich im grenzgebiet von wirtschaft und technik, von psychologie und ökologie und nicht zuletzt auch von kunst und kultur ab. der designer wird dabei zum anwalt des benutzers und arbeitet als dienstleister im auftrag des unternehmers.

im studium kann so ein vernetzter entwicklungsprozess beispielhaft in form von projektarbeiten erfahren werden. um dem umfassenden gestaltungsanspruch gerecht zu werden, gibt es am fh-studiengang in graz nicht nur eine designbetreuung durch professionelle gestalter, sondern auch eine begleitende betreuung durch experten aus den berufsfeldern für engineering, ergonomie, marketing, innovationsmanagement, kommunikation, darstellungstechniken und modellbau. dadurch wird die interdisziplinäre vorgangsweise im team bereits im studium trainiert.
die aufgabenstellung für die projektarbeiten erfolgt entweder in kooperation mit unternehmen oder studiengangintern. der ablauf gliedert sich in vier phasen:

phase 1: r e c h e r c h e / a n a l y s e
der einstieg erfolgt durch vielseitige recherchen zum gewählten thema (geschichtliche entwicklung, kulturelle bezüge, technologischer hintergrund, marktanalysen, usw.). anschliessend folgt eine intensive auseinandersetzung mit der wertewelt der zielgruppe (wie sieht sie aus? was liest sie? wie kleidet sie sich? wie ist ihr freizeitverhalten? welche produkte kauft sie?). die zielgruppendefinition erfolgt in form von collagen, sogenannten moodboards. den recherchen schliessen sich analysen an. auf grund der persönlichen schlussfolgerungen wird das briefing bzw. pflichtenheft ergänzt und mit wenigen schlagworten eine "botschaft" formuliert, die der entwurf später mit seiner produktsprache ausdrücken soll.

phase 2: k o n z e p t e
in der konzeptphase wird zuerst parallel produktsprachliche, gebrauchsorientierte und technische ideenfindung betrieben. die daraus abgeleiteten prinziplösungen werden anschliessend miteinander sinnvoll kombiniert. alternativen- und variantenbildung stehen im vordergrund. die darstellung erfolgt in form von von hand angefertigten skizzen bzw. renderings und arbeitsmodellen. anhand des selbst erarbeiteten briefings bzw. pflichtenheftes erfolgt die auswahl jenes konzepts, das für die weitere bearbeitung am erfolgversprechendsten ist.

in the reality of professional life, the design process is a very complex and holistic process of solutions to problems. it takes place in the border zone between economy and technology, psychology and ecology and, last but not least, art and culture. the designer becomes the user's advocate and works as a service provider for the companies.

while the student is studying, he/she can experience such a developmental networking process in exemplary fashion in the work they do on individual projects. in order to do justice to the extensive demands in design, the fh course of studies in graz offers not only design guidance through professional designers, but additional guidance by experts on engineering, ergonomics, marketing, innovative management, communications, presentation techniques, and model building. here, the student is already practicing interdisciplinary procedure within a team.
the task for the projects is set either in collaboration with companies or internally within the department. the procedure is divided into four phases:

phase 1: r e s e a r c h / a n a l y s i s
the first step is to do extensive research on the selected topic (historical development, cultural connections, technological background, market analysis, etc.). the next step is an intensive study of the target group (what does it look like? what does it read? how does it dress? how does it spend its leisure time? what kind of products does it buy?). the definition of the target group is done using collages, so-called moodboards. after the research is done, the analyses begin. the briefing/the assignment is supplemented based on the student's personal conclusions, and a concise "message" is formulated that the product language of the design is supposed to express.

phase 2: c o n c e p t s
in the conceptual phase, there is a parallel development of principle solutions, which pertain to product language, range of use, and technology, which are then combined to form a whole that makes sense. producing alternatives and different versions is of primary importance. the representation is done in the form of sketches and/or renderings done by hand and working models. based on the briefing/assignment, which the student has prepared, that concept, which seems to promise the most success in its further implementation, is selected.

designstudie: "pkw-anhänger", von julian hönig, leonard natterer, christian zwinger
design project: "passenger car trailer", by julian hönig, leonard natterer, christian zwinger

collage: wertewelt der zielgruppe "sportlich-ambitionierte erwachsene"
collage: values of the target group "grown-up sports enthusiasts"

phase 3: e n t w u r f
in der entwurfsphase erfolgen die optimierung des grundkonzeptes und die ausarbeitung aller details sowie die festlegung der masse, farben, materialien und oberflächen. die darstellung, meist in form von ansichten und schematischen schnitten, wechselt von der handzeichnung zur bearbeitung am computer – in der regel mit einem 3-d modeller. die virtuellen computermodelle können anschliessend zum teil direkt mit einer cnc-gesteuerten maschine aus hartschaum gefräst werden, zum teil werden werkstücke aber noch immer bewusst von hand hergestellt. wird nur mit virtuellen modellen gearbeitet, so ist zumindest ein proportionsmodell zur überprüfung der ergonomischen qualitäten und der "realen" proportionen – wenn immer möglich – im massstab 1:1 anzufertigen. im fahrzeugdesign wird ergänzend noch immer mit modellen aus clay (industrieplastillin) gearbeitet. von diesen werden negativformen abgenommen, die als basis für gfk-hartmodelle dienen.

phase 4: p r ä s e n t a t i o n
der abschluss der projektarbeit erfolgt durch eine sorgfältig vorbereitete präsentation der arbeitsergebnisse. die projektdokumentation muss den gesamten designprozess von der recherche bis zum designmodell beinhalten sowie eine klare, prägnante projektbeschreibung auf deutsch und englisch. wurde die projektarbeit in kooperation mit einem unternehmen durchgeführt, so sind die firmenvertreter bei der präsentation anwesend und geben ihren kommentar zu den einzelnen projekten ab.

allgemeine b e u r t e i l u n g s k r i t e r i e n für design-studien/projektarbeiten:
1 praktische funktion: handhabung, gebrauch, transport, lagerung, . . .
2 ästhetische funktion: formale qualität, proportionen, farbe, oberflächen, . . .
3 symbolische funktion: übereinstimmung der produktsprache mit der zielgruppe, der selbstverfassten "botschaft" und gegebenenfalls mit der corporate identity eines unternehmens, . . .
4 anzeichenfunktion: selbsterklärung im gebrauch, ...
5 technisch/wirtschaftliche funktionen: herstellbarkeit, unterhaltskosten, . . .
6 ökologische funktionen: energieverbrauch, materialwahl, recycling, . . .
7 präsentation: argumentationsstärke bei der persönlichen vorstellung, aussagekraft der dokumentation, . . .

phase 3: d e s i g n d r a f t
in the design phase, the basic concept is optimized, and the details such as measurements, colors, materials, and surfaces are determined. the descriptive representation, usually in the form of views and schematic sectional models now switches from sketches by hand to work on the computer, usually a 3-d modeller. the virtual computer models can then be, at least in part, directly cut out of hard foam by a cnc-controlled machine. sometimes, the pieces are still deliberately made by hand. if only virtual models are being worked on, at the very least, a 1:1 (if at all possible) proportional model should be prepared, so that ergonomic qualities and the "real" proportions can be checked out. clay (industrial plasticine) models whose negative forms represent the basis of grp models are still additionally used in vehicle design.

phase 4: p r e s e n t a t i o n
the final step of the project is a carefully prepared presentation of the results. the project documentation must include the entire design process from the research to the design model, as well as a clear, precise project description in german and english. if the project was prepared in collaboration with a company, the company representatives are at the presentation and comment on the individual projects.

general e v a l u a t i o n c r i t e r i a for design-studies/projects:
1 practical function: handling, use, transport, storage, . . .
2 esthetic function: form, proportions, color, surfaces, . . .
3 symbolic function: coordination of the product language with the target group, the "message" the student has written, and, if appropriate, with the corporate identity of the company, . . .
4 signal function: self-explanatory use, . . .
5 technical/economic functions: producibility, maintenance costs, . . .
6 ecological functions: energy consumption, choice of material, recycling, . . .
7 presentation: strength of argumentation in the personal presentation, informative value of the documentation, . . .

inhalt / *contents:*

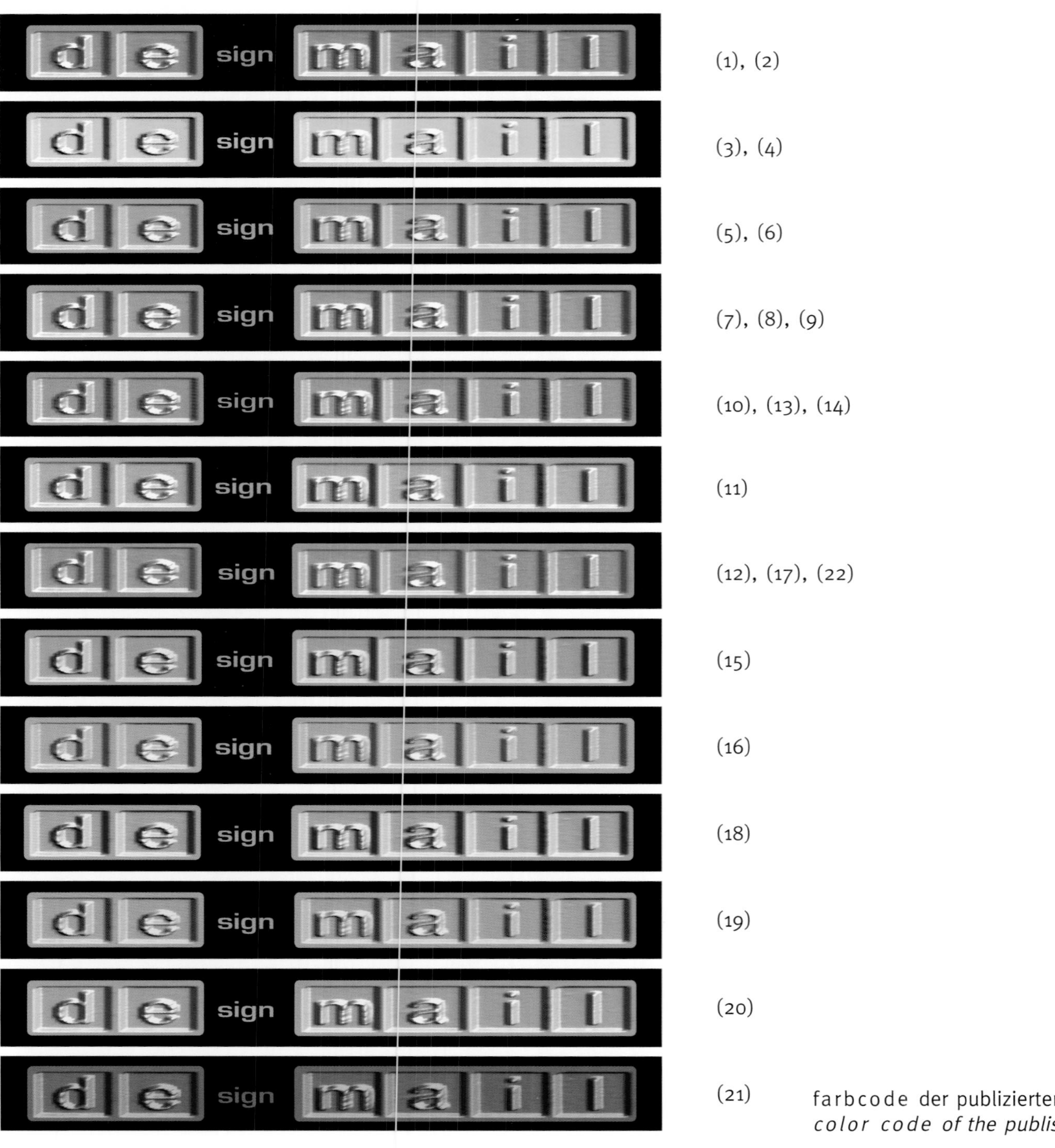

..........14

(1), (2)

(3), (4)

(5), (6)

(7), (8), (9)

(10), (13), (14)

(11)

(12), (17), (22)

(15)

(16)

(18)

(19)

(20)

(21)

farbcode der publizierten arbeiten mit projektnummer
color code of the published projects including number

projekt: » trinkwasserbrunnen « - geladener wettbewerb der stadt graz
jury: dirk schmauser (porsche design), sigmar willnauer (doods tm)
3. semester/winter 1996/97

betreuer: gerhard h e u f l e r - gerald k i s k a - professoren für design -
georg w a g n e r - professor für engineering -
werner k l e i s s n e r - modellbau -

thema: stadtbewohner und stadtbesucher möchten im vorübergehen einen schluck wasser trinken, ihr obst waschen oder auch nur den kopf unter das wasser halten, weil es so schwül ist. aufgabe ist es, auf basis eines technikmoduls einen trinkwasserbrunnen zu entwerfen, der als "grazer brunnen" zeichenfunktion hat und einen beitrag zur städtischen lebensqualität leistet. das siegerprojekt wurde realisiert und in mehreren exemplaren in graz aufgestellt.

project: » drinking fountain « competition by invitation of the city of graz
jury: dirk schmauser (porsche design), sigmar willnauer (doods tm)
3rd semester/fall 1996/97

advisors: gerhard h e u f l e r - gerald k i s k a - professors of design -
georg w a g n e r - professor of engineering -
werner k l e i s s n e r - model building -

subject: city residents and visitors sometimes want a drink of water, or want to wash a piece of fruit, or even just hold their head under water because it is so humid. the task is to design a drinking fountain based on a technological module that would function as a landmark – the "graz fountain" – and would be a welcome addition to city quality of life. the winning project was realized and several fountains were set up in graz.

trinkwasser b

sign

p r o j e k t a r b e i t e n d e s f h - s t u d i e n g a n g e s i n d u s t r i a l d e s i g n - g r a z

p r o j e c t s o f t h e f h - d e g r e e c o u r s e i n d u s t r i a l d e s i g n - g r a z

1996/97
(1)

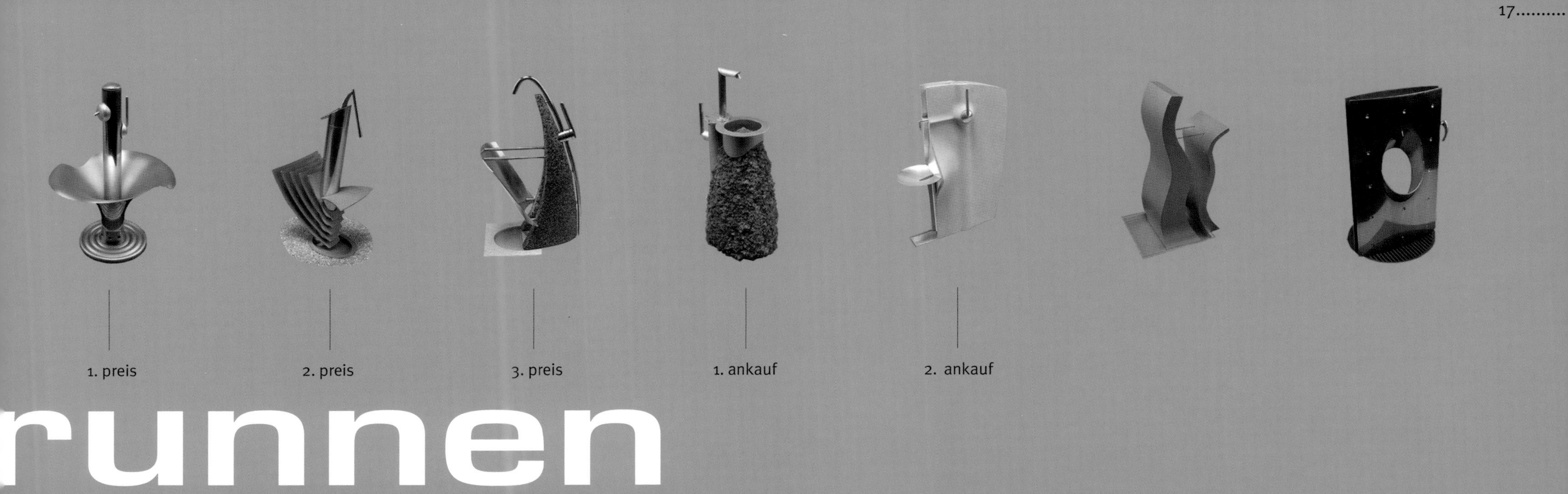

1. preis

2. preis

3. preis

1. ankauf

2. ankauf

runnen

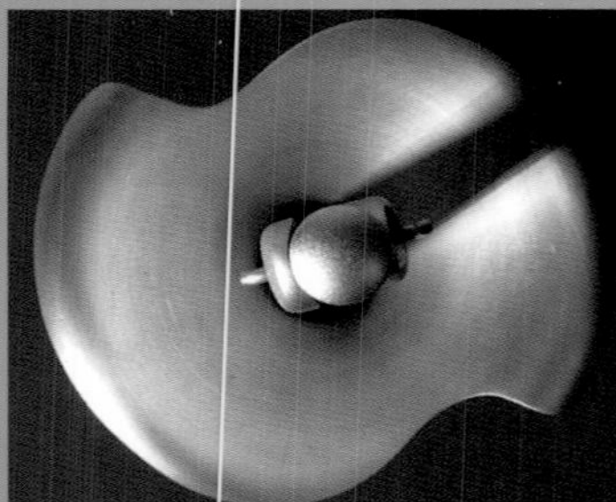

» 1. preis «
» niki pelzl «

--- jury: die vergabe des 1. preises durch die jury und die damit verbundene realisierung des brunnens in der stadt graz begründet sich in erster linie dadurch, dass der entwurf klassische elemente des archetypischen brunnens und zeitgenössisches formen- und materialverständnis erfolgreich in einem lyrischen ergebnis vereint ---

--- jury: the 1st prize was awarded by the jury and, subsequently, the fountain was realized in graz primarily because the design successfully combines the classical elements of the archetypal fountain with a contemporary understanding of forms and materials to achieve a lyrical result ---

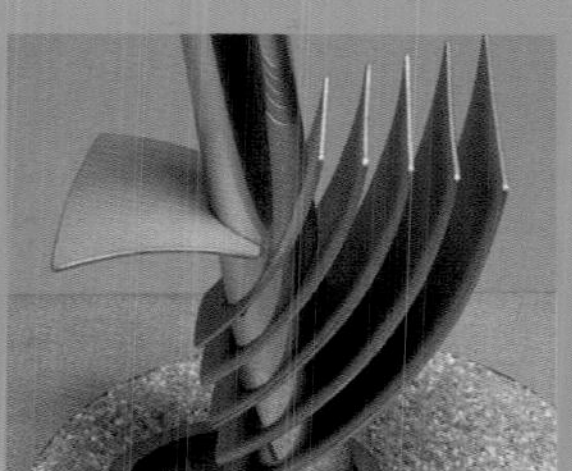

» 2. preis «
» christian winkler «

--- jury: der entwurf zeichnet sich durch einen sehr hohen zeichencharakter und prägnanz der form aus -- das design wird von einer präzisen detail-entwicklung bestimmt ---

--- jury: the design is distinguished both by its rich symbolism and its striking form -- the design is defined by a precise development of detail ---

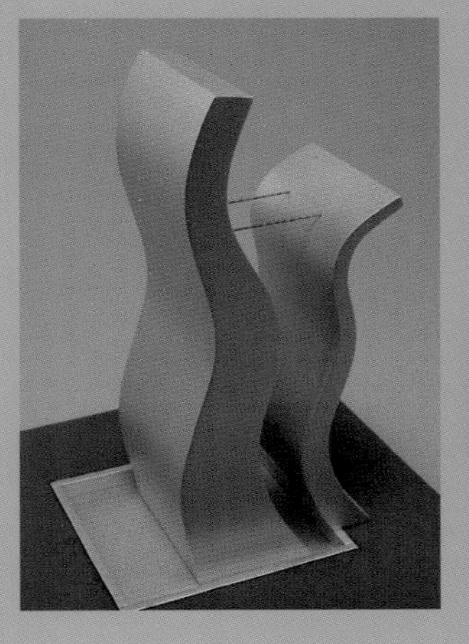

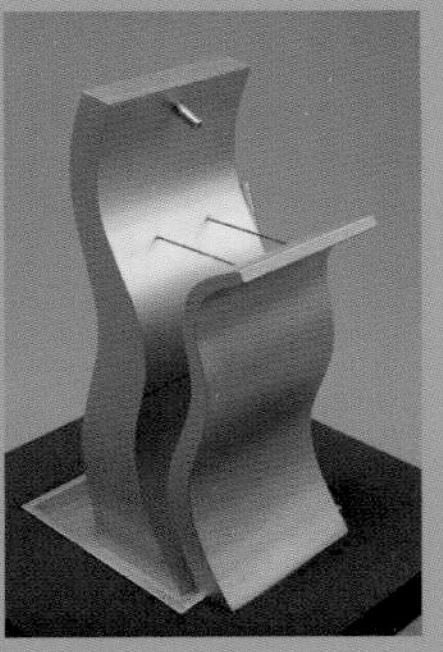

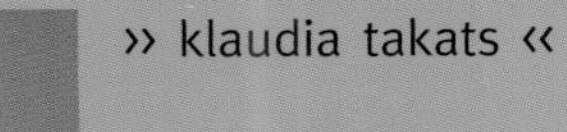

» klaudia takats «

» harald auer «

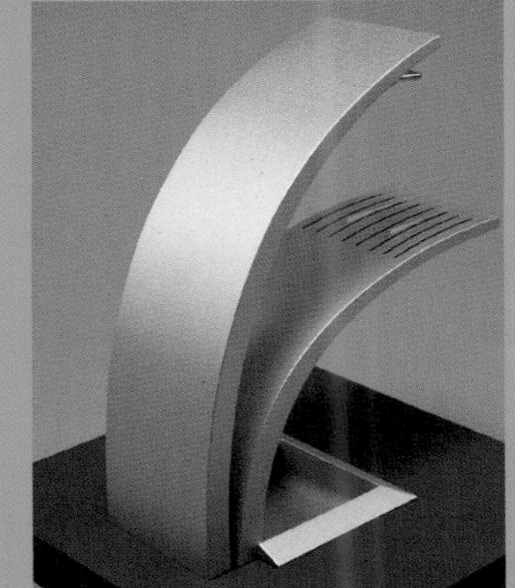

» sylvia feichtinger «

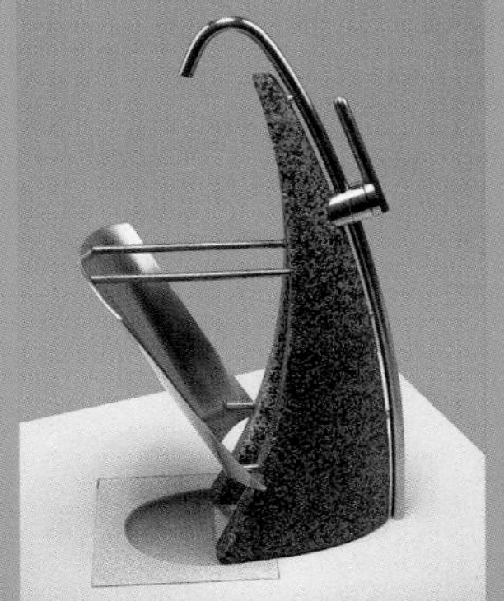

» 3. preis «

» david stelzer «

--- jury: die bewegung und dynamik des wassers wurden in diesem entwurf sichtbar nachvollzogen -- im städtischen raum setzt dieser brunnen ein unübersehbares zeichen ---

--- jury: in this design, the movement and dynamics of water were visibly reproduced -- this fountain sets a highly visible example in inner city space ---

» 2. ankauf «
» michael tropper «

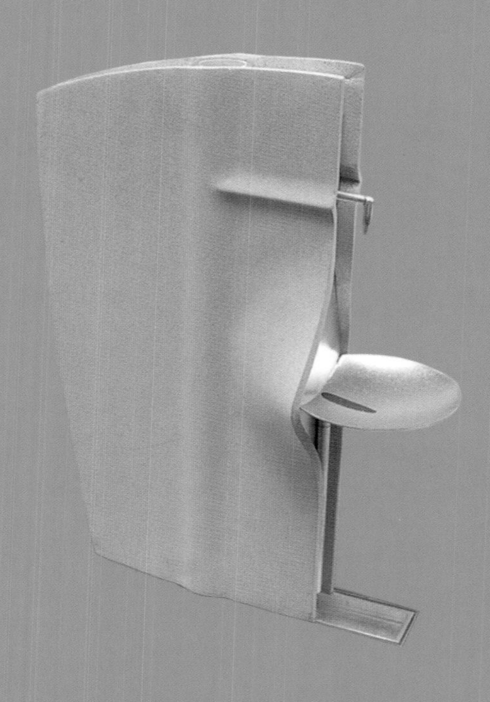

--- jury: durch die reliefartige modellierung der sichtbetonplatten werden die inneren funktionen sichtbar gemacht -- als stele wird dem brunnen eine hohe zeichenhaftigkeit zugesprochen -- analogieschlüsse zur spaltförmigen erweiterung einer quelle im felsmassiv ---

--- jury: the inner functions are made visible by means of the relief-like shaping of the exposed concrete slabs -- the upright slab lends the fountain a great deal of symbolic significance -- suggests an analogy with the widening crevice in solid rock created by a spring ---

» 1. ankauf «
» constantin luser «

--- jury: die kontrapunktik der textur, die glatte und exakte kälte des stahlzylinders sowie die grobe rohrbauweise wird als versuch gewertet, die ambivalente haltung des urbanen bewohners zur natur und technik zu beschreiben -- material stein: tuffstein, wasserspeichernde wirkung, begünstigt die bildung eines grünmilieus -- material metall: nirosta, oberfläche sandgestrahlt (pflegeleicht und langlebig) ---

--- jury: the counterpoint-like qualities of the texture, the smooth and precise coldness of the steel cylinder, as well as the rough construction using a pipe is considered to be an attempt to describe the ambivalent attitude of the urban resident to nature and technology -- stone material: tuff stone, accumulates and retains water, benefits the cultivation of a green environment -- metal material: stainless steel, surface sand-blasted (low-maintenance and durable) ---

» tanja langgner «

» gernot puschner «

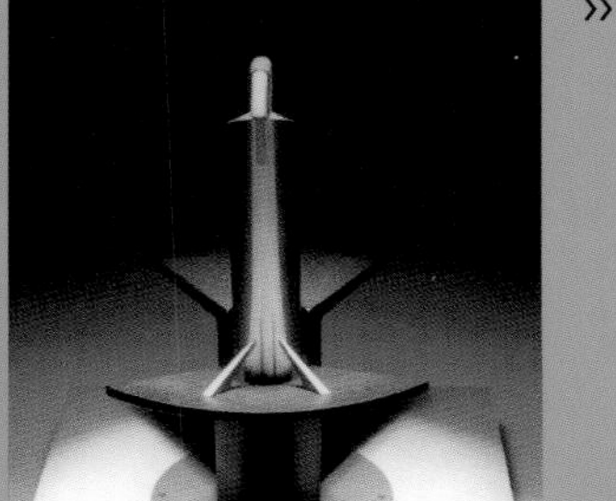

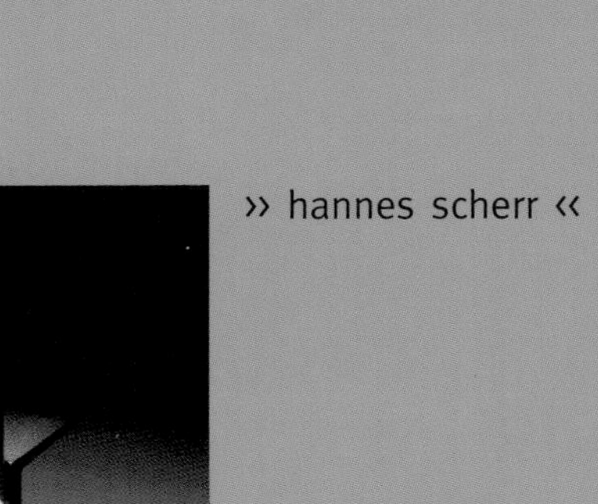

» hannes scherr «

» silke spath «

projekt: » digitale kamera «
- in kooperation mit sony
4. semester/sommer 1997

betreuer: gerhard h e u f l e r - gerald k i s k a - professoren für design -
georg w a g n e r - professor für engineering -
werner k l e i s s n e r - modellbau -

thema: auf der technischen basis von sony – mit einem visionshorizont von fünf jahren – sollen design-studien entwickelt werden, die den speziellen anforderungen von selbstdefinierten zielgruppen entsprechen. die für die jeweilige zielgruppe entwickelten features sowie der einsatzbereich sollen durch die gestaltung funktional umgesetzt und entsprechend visualisiert werden. zukünftige technische verbesserungen können eingebracht werden.

project: *» digital camera «*
- in collaboration with sony
4th semester/spring 1997

advisors: *gerhard* *h e u f l e r* - gerald k i s k a - *professors of design -*
georg *w a g n e r - professor of engineering -*
werner *k l e i s s n e r - model building -*

subject: *planned for actual realization five years down the road, design studies based on sony technology will be developed, which are suited for special requirements of self-defined target groups. the features developed for each target group, as well as the deployment sector will be appropriately visualized and functionally implemented through the design. future technical improvements can be integrated later on.*

my first sony

sony image

digitale kamera

sign

projektarbeiten des fh-studienganges industrial design-graz

projects of the fh-degree course industrial design-graz

1997
(2)

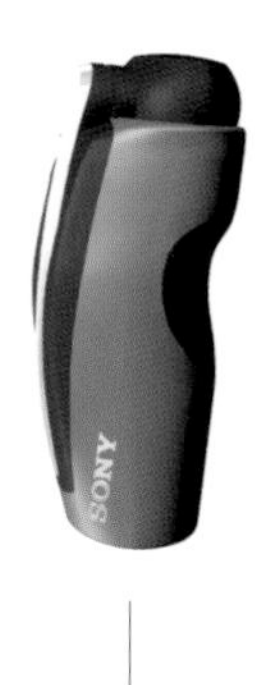

3rd eye

always a snap

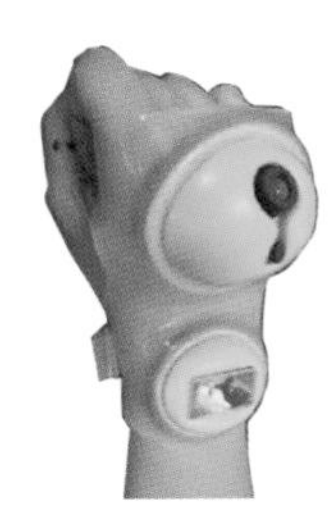

jelly fish

slim autarky

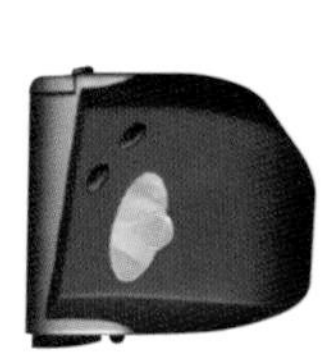

marilyn

travelmate

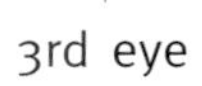

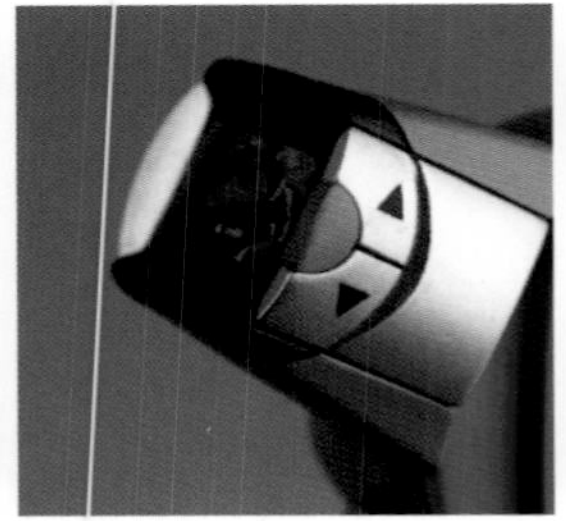
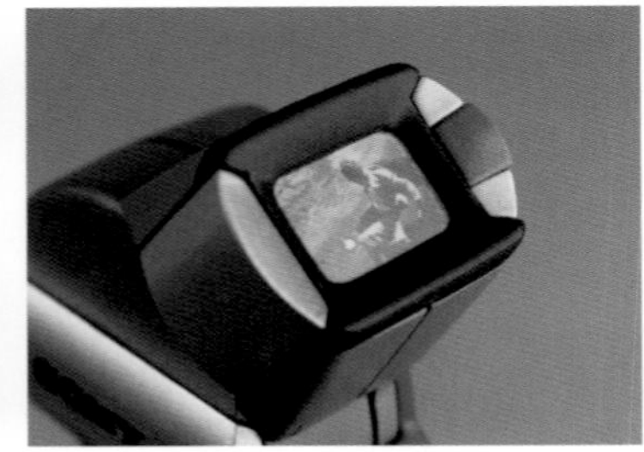

» always a snap «
» johannes scherr «

--- eine digitale kamera als begleiter in allen lebenssituationen: im gebrauch legt das griffelement das drehbare objektiv und das zoomrad frei -- zielgrupppe: - designbewusste jüngere menschen ---

--- a digital camera as a companion in any life situation: in normal use, the handle opens up the swiveling lens and the zoom wheel -- target group: design-conscious people ---

--- kombination aus digitaler kamera und digitalem diktafon -- extrem flaches gehäuse für die jackentasche -- schwenkbares objektiv -- ausklappbares lc-display -- zielgruppe: - businessman/-woman für beruf und freizeit ---

--- combination of digital camera and digital dictation machine -- extremely flat case that has room in a jacket pocket -- pivoting lens -- fold-out lc display -- target group: the business-person on the job and for personal use ---

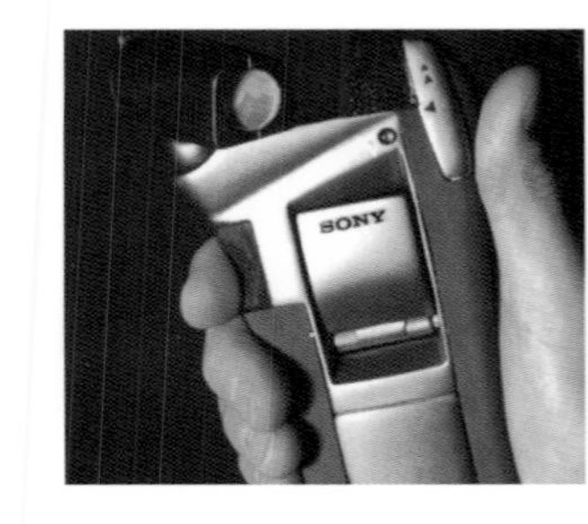

» slim autarky «
» harald auer «

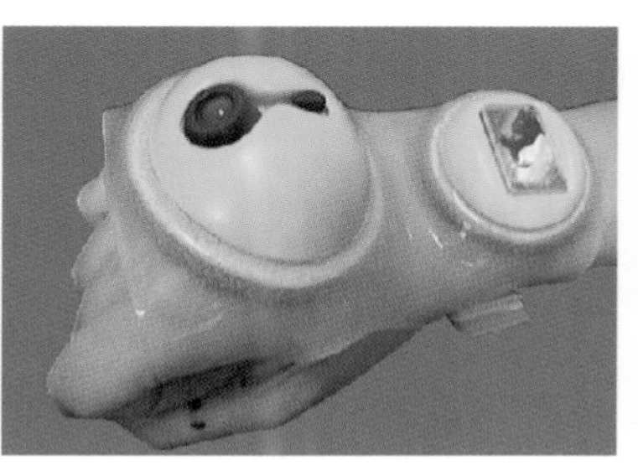

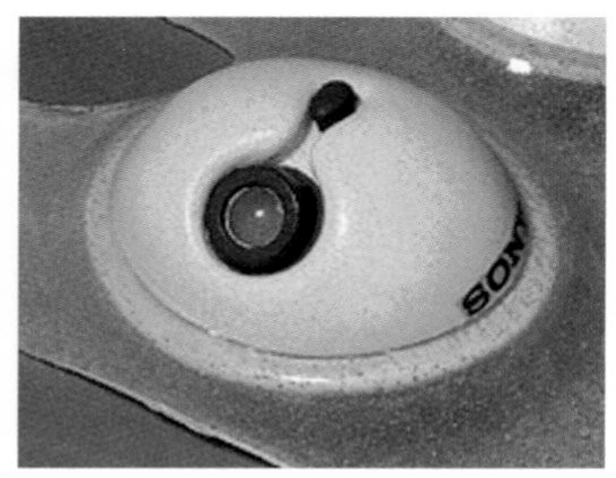

» jelly fish »

» claudia takats »

--- zielgruppe: - wassersportarten wie windsurfen, wellenreiten... -- material: transparent, soft, robust, seewasserfest, leicht... ---

--- target group: water sports such as windsurfing, surfing ... -- material: transparent, soft, robust, saltwater-proof, light ... ---

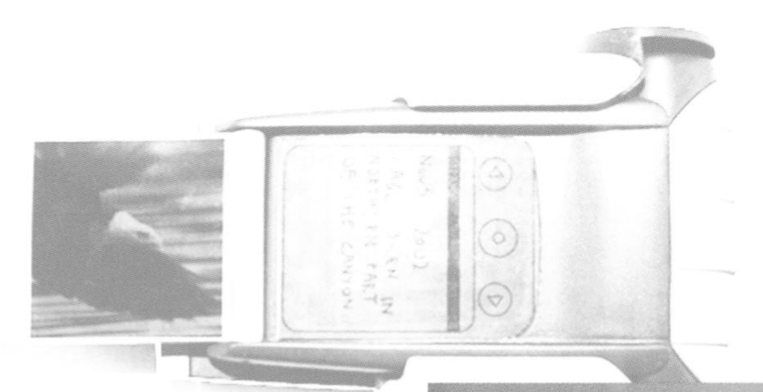

» travelmate «

» michael tropper «

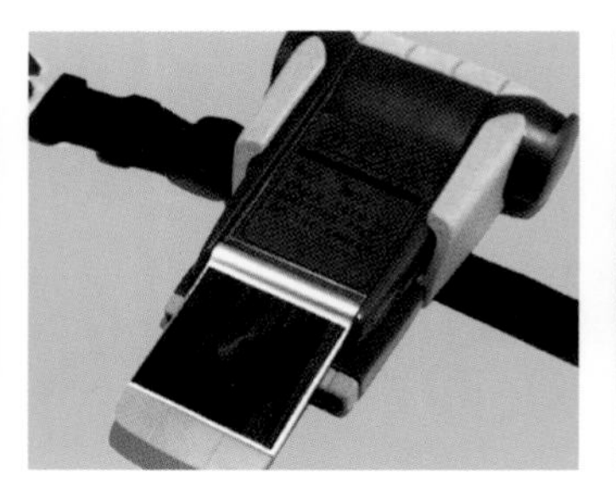

--- zielgruppe: - expeditionen, - forschungsreisende, - tramper, - alpin trekking -- kombination aus: - digitaler kamera, - digitalem reisetagebuch, - gps und höhenmesser ---

--- target group: - expeditions, - explorers and researchers, - hitchhikers, - alpine trekking -- combination of: - digital camera, - digital travel diary, - gps and altimeter ---

» 3rd eye «
» david stelzer «

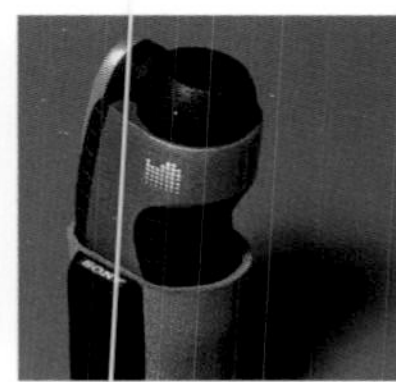

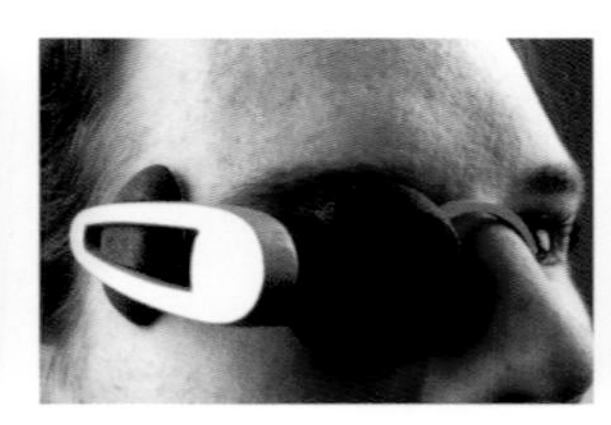

--- die "3rd eye" wurde speziell für die ansprüche des fotoreporters konzipiert -- so bietet sich die möglichkeit, den sucher abzunehmen und wie eine brille zu verwenden - - somit hat man das sucherbild direkt vor auge, egal, wohin die kamera gerichtet wird (überkopf, ums eck, nach hinten) ---

--- the "3rd eye" was designed especially for the photojournalist's needs -- it offers the possibility of removing the viewfinder and using it as eyeglasses -- this way you always have the viewfinder image directly in front of you, no matter which direction the camera is pointing in (over the head, around the corner, backwards) ---

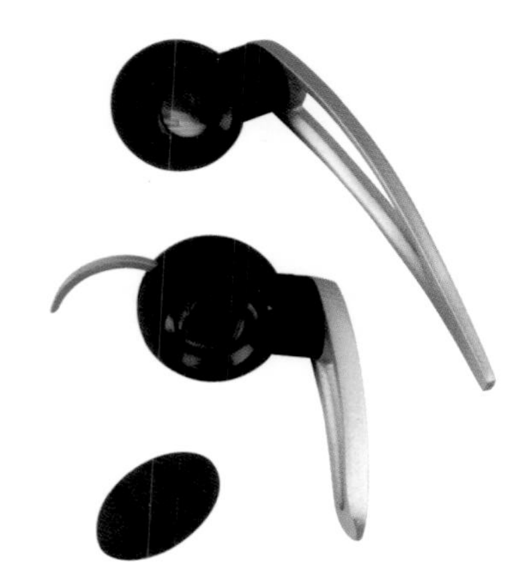

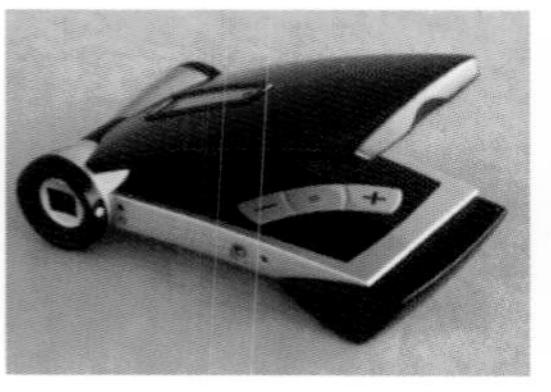

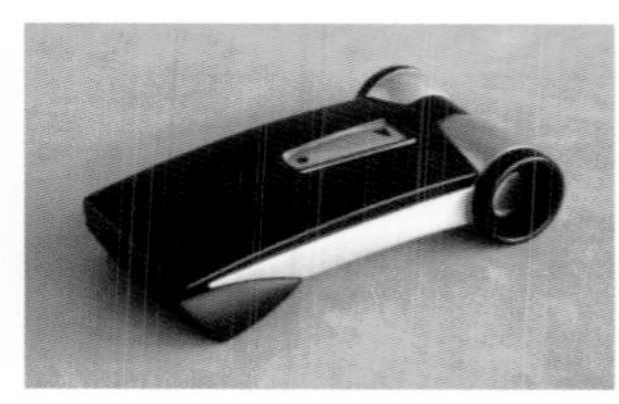

» sony image «
» christian winkler «

--- zielgruppe: die "image" möchte die generation der anfang 30-jährigen ansprechen: - erfolg im beruf, - ausgeprägtes qualitätsbewusstsein -- ausformung, gewicht und grösse ermöglichen es, die kamera in der tasche eines kleidungstücks griffbereit am körper zu transportieren ---

--- target group: the generation that is now in its early 30's: - professional success, - definite awareness of quality -- form, weight and size make it possible to carry the camera in a pocket and always have it available ---

--- zielgruppe: - die selbstbewusste lady um 50 -- elegantes erscheinungsbild -- liegt perfekt in der hand -- passt in jede tasche ---

--- target group: - the confident lady in her 50's -- elegant appearance -- convenient to hold -- fits into any purse ---

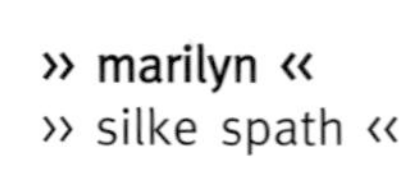

» marilyn «
» silke spath «

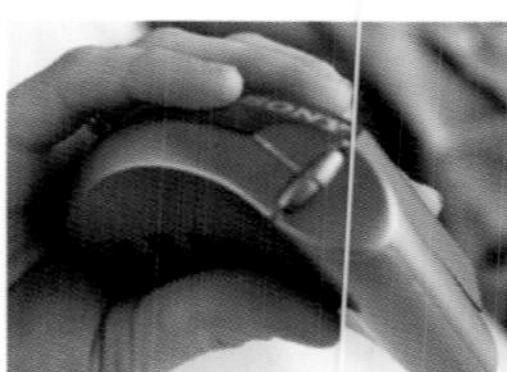

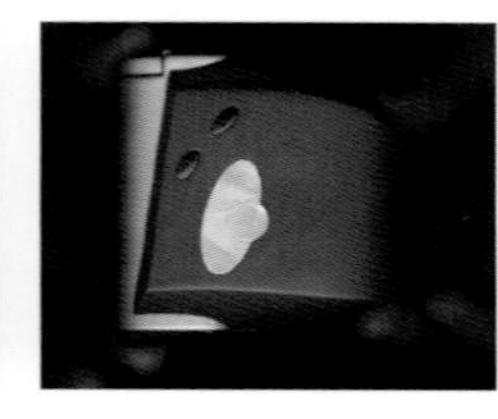

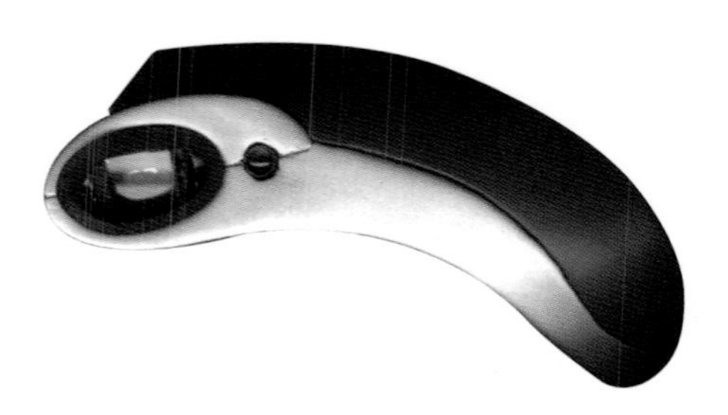

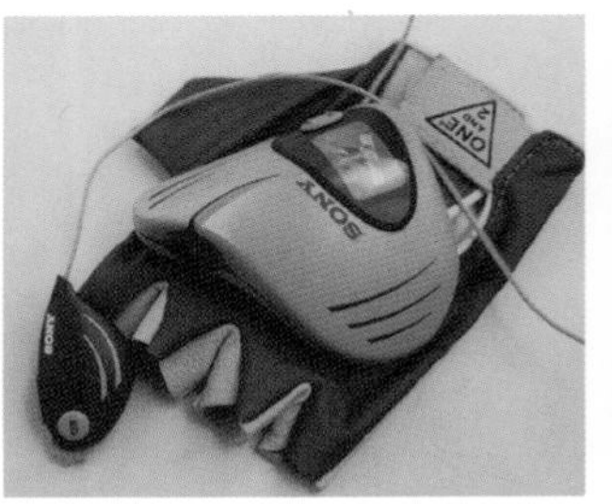

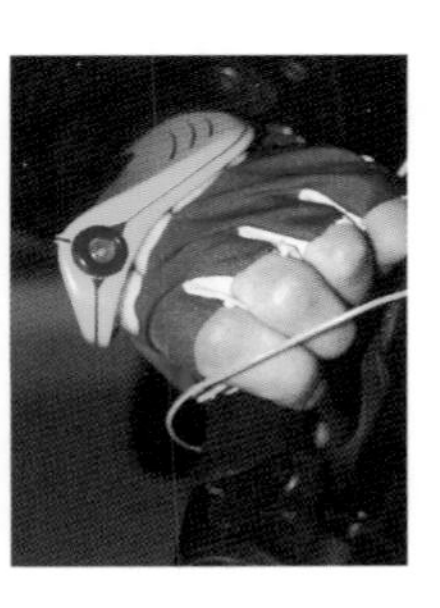

» xtreme «

» sylvia feichtinger «

--- zielgruppe: - mountainbiking, - snowboarding, - bungyjumping etc... -- man benutzt die kamera mit oder ohne handschuh; sie ist durch einen an der unterseite montierten klettverschluss an anderen körperstellen wie z.b. schulter, oberschenkel zu befestigen ---

--- target group: - mountain biking, - snowboarding, - bungy jumping, etc... -- you can use the camera with or without gloves; a velcro strip on the bottom of the camera makes it possible to attach it to other parts of the body such as shoulder or thigh ---

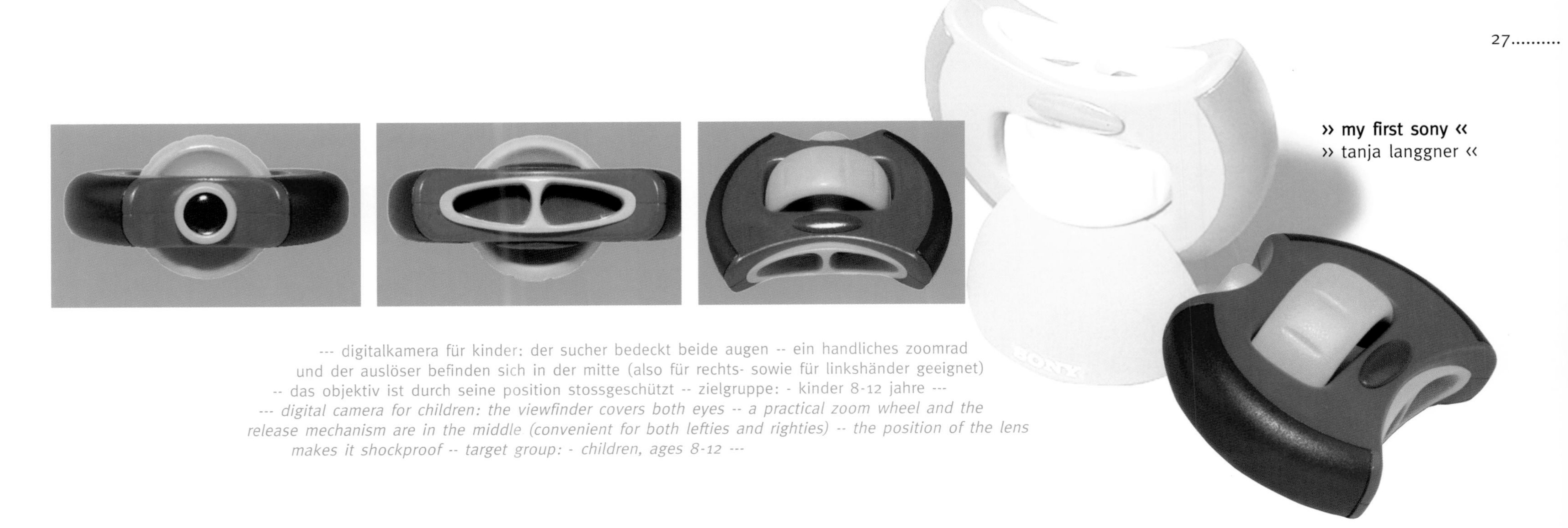

» my first sony «

» tanja langgner «

--- digitalkamera für kinder: der sucher bedeckt beide augen -- ein handliches zoomrad und der auslöser befinden sich in der mitte (also für rechts- sowie für linkshänder geeignet) -- das objektiv ist durch seine position stossgeschützt -- zielgruppe: - kinder 8-12 jahre ---

--- digital camera for children: the viewfinder covers both eyes -- a practical zoom wheel and the release mechanism are in the middle (convenient for both lefties and righties) -- the position of the lens makes it shockproof -- target group: - children, ages 8-12 ---

FH JOANNEUM
FACHHOCHSCHULSTUDIENGÄNGE DER TECHNIKUM JOANNEUM GMBH

1997/98 (3)

projekt: » lawinen pieps «
- in kooperation mit puls elektronik + seidel elektronik
3. semester/winter 1997/98

betreuer: gerhard h e u f l e r - professor für design -
georg w a g n e r - professor für engineering -
werner k l e i s s n e r - modellbau -

thema: die suche nach lawinenverschütteten geschieht unter grossem stress. es sollen geräte konzipiert werden, die auf möglichst einfache weise die retter schnell zum verschütteten führen. ergonomie und anzeichenfunktion stehen dabei im vordergrund. als zielgruppen kommen neben klassischen tourenfahrern auch snowboarder, bergprofis der alpinrettung oder hüttenwirte in frage. visionshorizont: fünf jahre.

project: » avalanche beeper «
- in collaboration with puls elektronik + seidel elektronik
3rd semester/fall 1997/98

advisors: gerhard h e u f l e r - professor of design -
georg w a g n e r - professor of engineering -
werner k l e i s s n e r - model building -

subject: a search for people buried by avalanches is always a time of great stress. this project plans to design devices, which will bring the rescuers to the victims as quickly and as easily as possible. ergonomics and signal function are an extremely important consideration. besides the classic long-distance skiers, other target groups are snowboarders, mountaineers in alpine rescue organizations and mountain lodge owners. realization planned within five years.

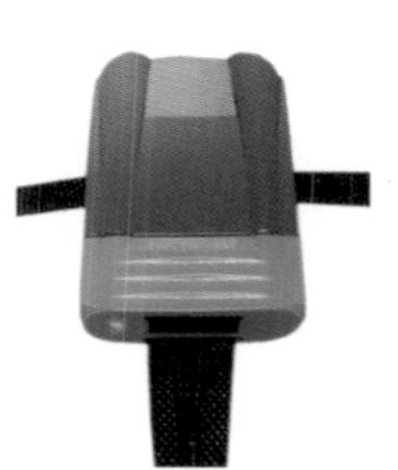

g2

digital-pieps

lawinen pieps

sign

projektarbeiten des fh-studienganges industrial design-graz
projects of the fh-degree course industrial design-graz

1997/98
(3)

pieps

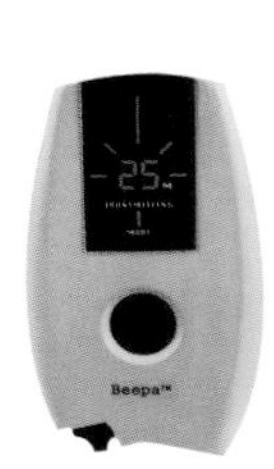

beepa

scout-x

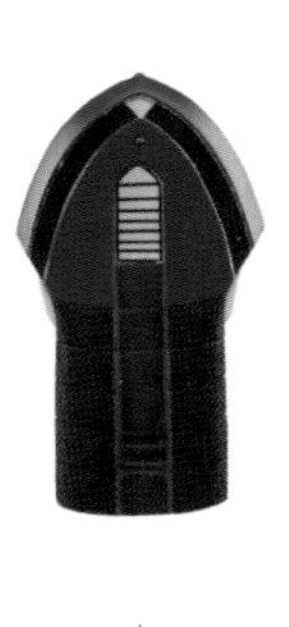

pinpoint

rupie

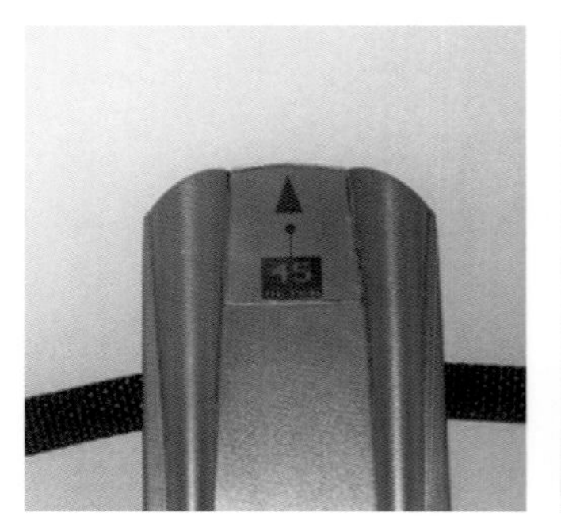

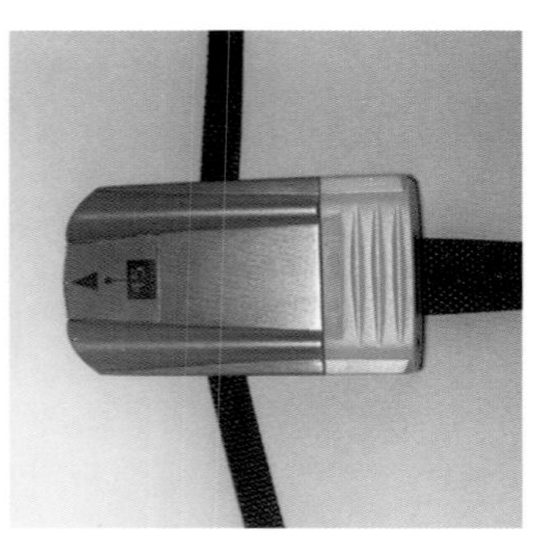

» g2 «
» christian becker «

--- suchmodus -- falsche richtung:
pfeil gibt richtungskorrektur an --
anzeige signalisiert entfernung in metern --
--- search mode -- wrong direction: arrow shows directional correction -- display shows distance in meters --

-- suchmodus -- richtung gefunden:
pfeil gross nach vorne --
anzeige signalisiert entfernung in metern ---
-- search mode -- correct direction found: large arrow pointing forwards -- display shows distance in meters --

-- wegen nachlawinen:
nach ca. 3 minuten aufforderung
zur bestätigung des suchmodus -- sonst schaltet
das gerät automatisch auf senden um ---
-- because of subsequent avalanches: after 3 minutes confirm search mode -- otherwise the device automatically switches to send --

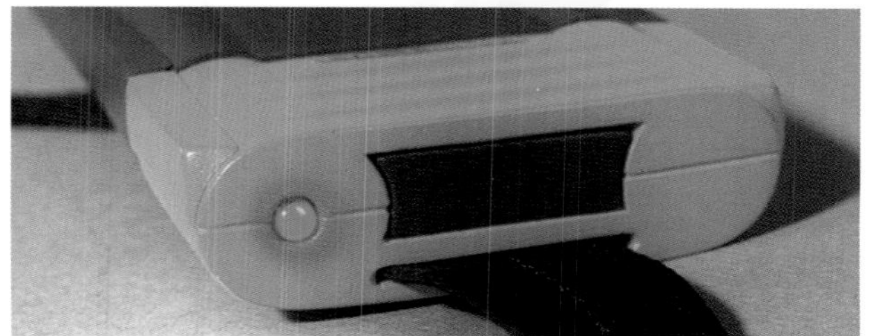

-- umschalten von senden auf suchen
durch ziehen des unteren gehäuseteiles ---
*-- switching from send to search
by pulling the lower part of the casing ---*

--- ergonomie- und formstudien anhand
von hartschaummodellen führen zu einer strukturierten greif-
zone -- unmissverständliche anzeige der richtung durch pfeilsymbolik --
umschalten auf empfang durch betätigung des drucktasters ---

*--- ergonomics and form studies on models made of hard
foam make a structured gripping zone possible -- arrows unmistak-
ably display the direction -- switching over to receiving by activating
a pushbutton ---*

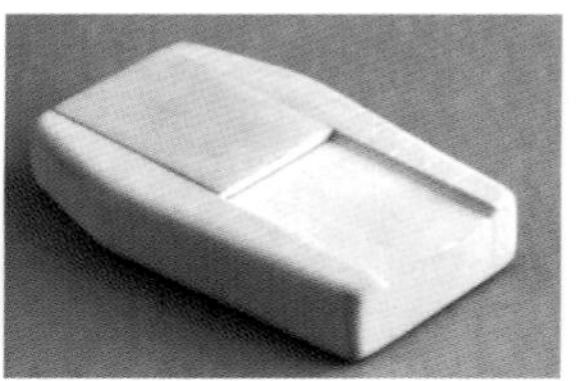

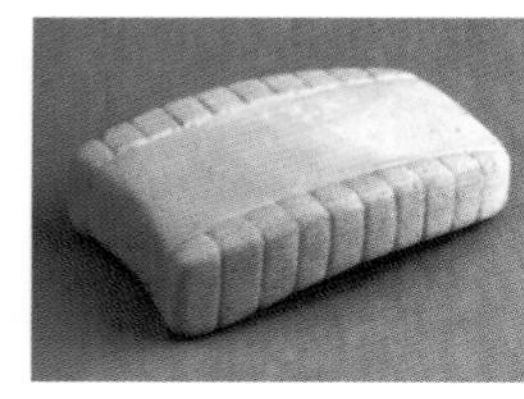

» scout-x «
» karen rosenkranz «

--- "rupie" wurde für den professionellen alpinen bereich entwickelt --
es verfügt über eine seilzugaufhängung, die eine einwandfreie bedienung durch wegziehen vom
körper ermöglicht und trotzdem eine schnelle rückführung des gerätes in seine aus-
gangsposition am körper durch federkraft sichert ---

*--- "rupie" was developed for professional alpinists -- it has a cable pull suspension device that
works easily by pulling it away from the body and assures a quick return of the device to its original
position on the body by means of a spring ---*

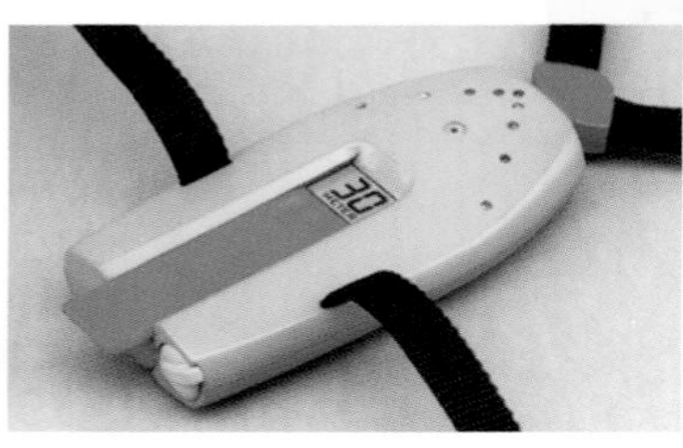

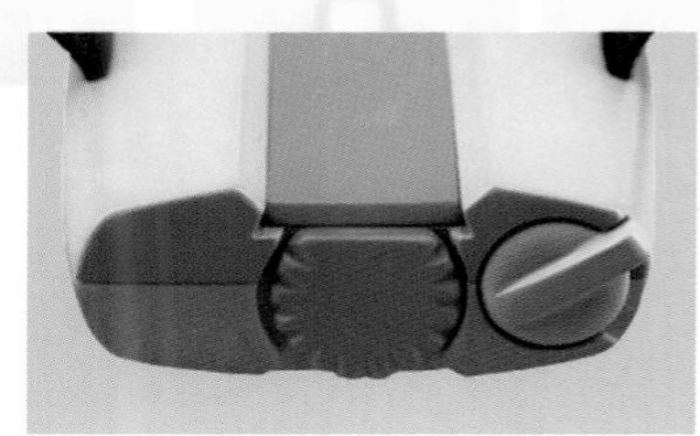

» rupie «
» rudolf steiner «

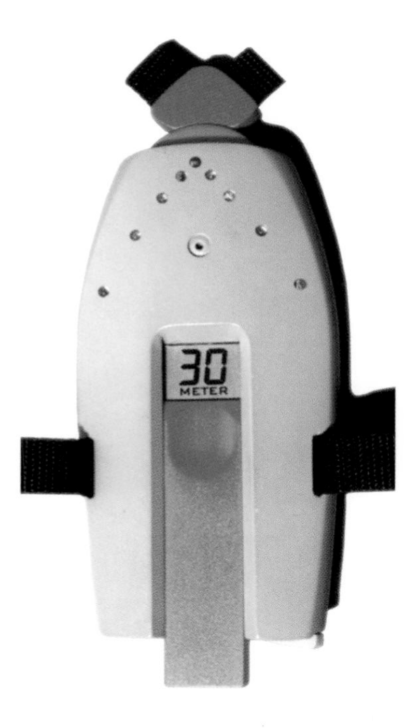

» digital-pieps «
» julian hönig «

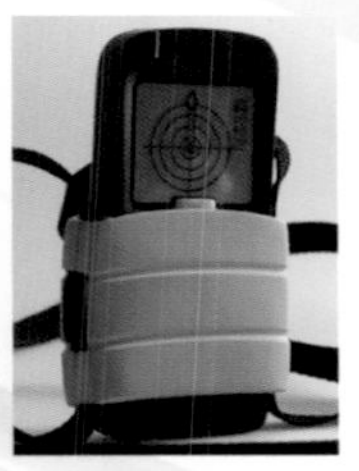

--- multiple searching d.h. ortung mehrerer verschütteter -- optische anzeige über lcd (radarähnlicher aufbau) -- einfache handhabung -- aktivierung durch verschiebung der lcd-abdeckung ---
--- multiple searching i.e. finding the location of more than one person buried by the avalanche -- optical lcd display (radar-like construction) -- easy handling -- activation by shifting the lcd cover ---

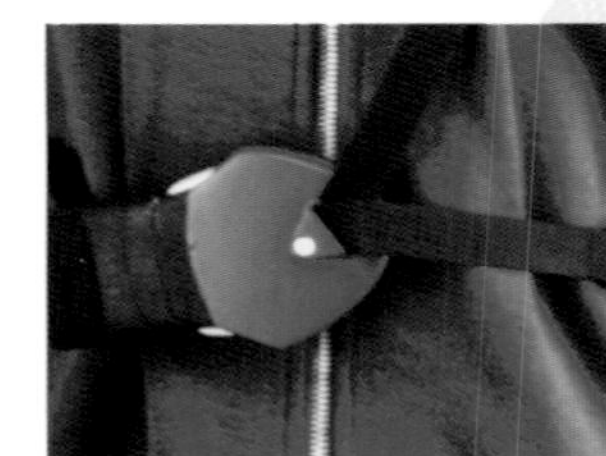

» pinpoint «
» valerie trauttmansdorff «

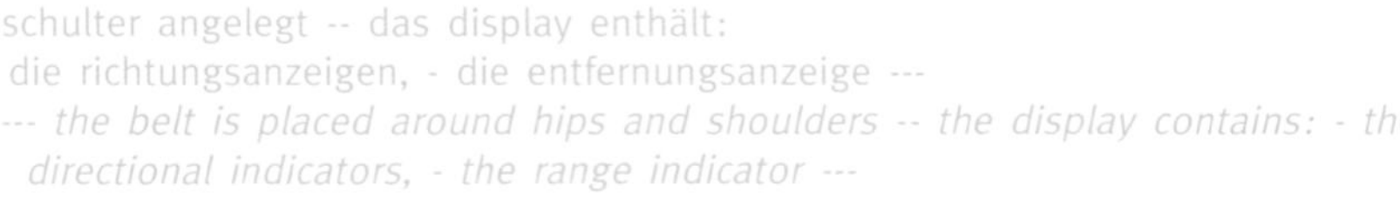

--- der gurt wird um hüfte und schulter angelegt -- das display enthält: - die richtungsanzeigen, - die entfernungsanzeige ---
--- the belt is placed around hips and shoulders -- the display contains: - the directional indicators, - the range indicator ---

» beepa «
» roland lindner «

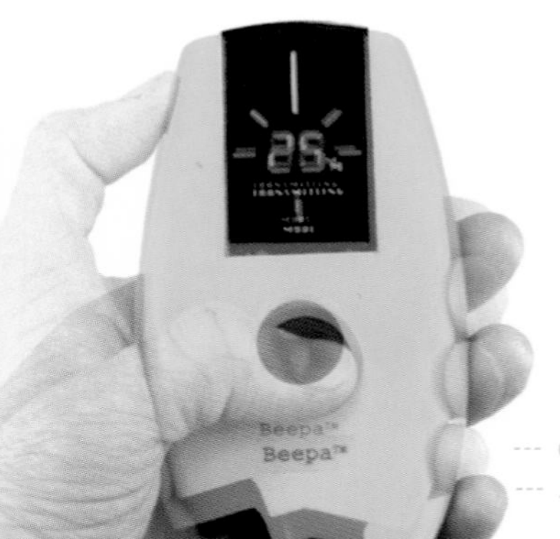

--- einfaches umschalten, senden/suchen ---
--- simple switching send/search ---

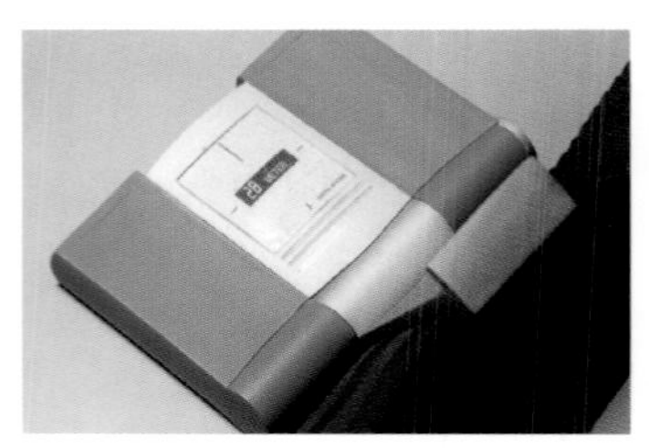

» pieps «
» martin wöls «

--- mit dem aufklappen des gerätes aktiviert sich das display und richtungs- sowie entfernungsanzeige erscheinen ---
--- opening the device activates the display and both the directional and range indicators appear ---

mobile leuchten

1997/98 (4)

projekt: » mobile leuchten «
4. semester / sommer 1997

betreuer: gerald k i s k a - professor für design -
georg w a g n e r - professor für engineering -
werner k l e i s s n e r - modellbau -

project: *» mobile lamps «*
4th semester/spring 1997

advisors: *gerald k i s k a - professor of design -*
georg w a g n e r - professor of engineering -
werner k l e i s s n e r - model building -

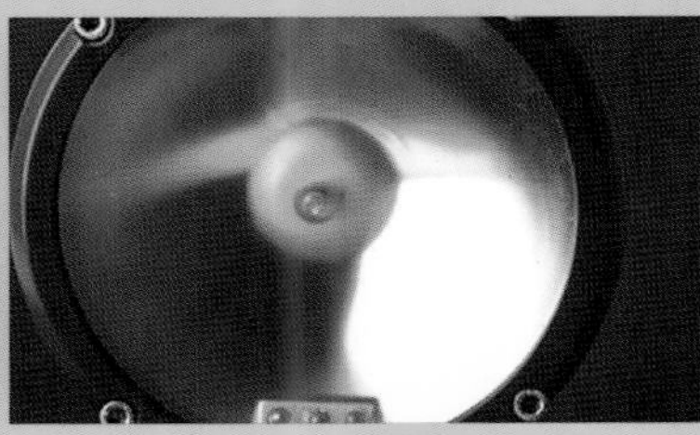

» feuerwehrlicht *fire-brigade lighting* «
» leonard natterer «

--- die leuchte wird im einsatzfall mitgeführt, um den arbeitsbereich optimal auszuleuchten (z.b. eine bergung bei nacht) -- abgestellt am griff, geschwenkt und ausgerichtet -- gleichzeitig kann mit einer hand der lichtkegel am fokusrad optimal eingestellt werden -- ein blinkmodus ist ebenfalls teil der funktionen -- das gelenk ist in zwei extrempunkten sperrbar ---

--- the lamp is brought along in emergencies for optimal lighting in the area of operations (e.g. rescue work at night) -- propped up on the handle, pivoted and aligned -- at the same time, the light cone can be optimally adjusted on the focus wheel with one hand -- blink mode is also available -- the joint can be locked in two extreme positions ---

--- beleuchtung für skater -- ausübung dieses sports auch nach einbruch der dunkelheit -- das licht am helm -- justierbarer lichtkegel nach vorne und integrierte rücklichter nach hinten (2 leuchtdioden) -- ausdruck von bewegung und geschwindigkeit ---

--- lighting for skaters -- facilitates practicing these sports in the dark -- light on the helmet -- adjustable light cone toward the front and integrates rear lights toward the back (2 light diodes) -- signals movement and speed ---

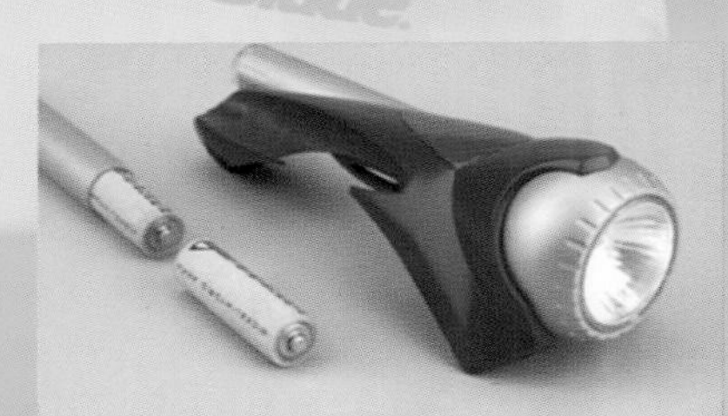

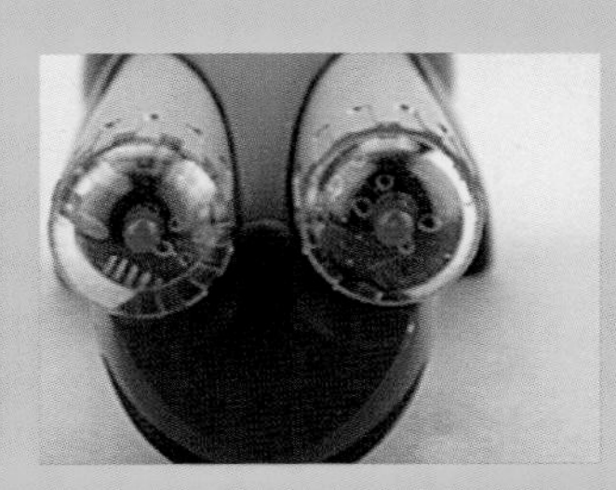

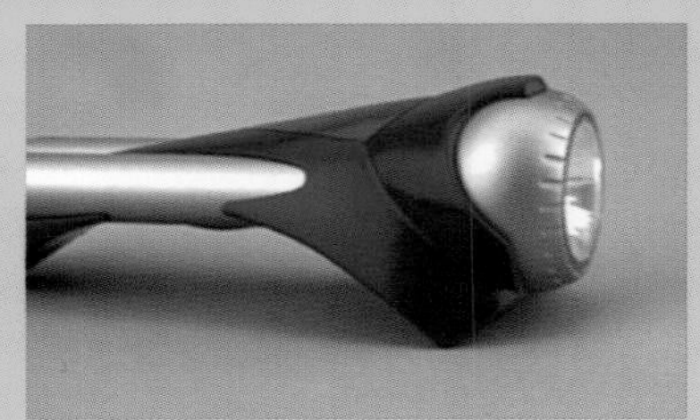

» skaterlicht «
» thomas binder «

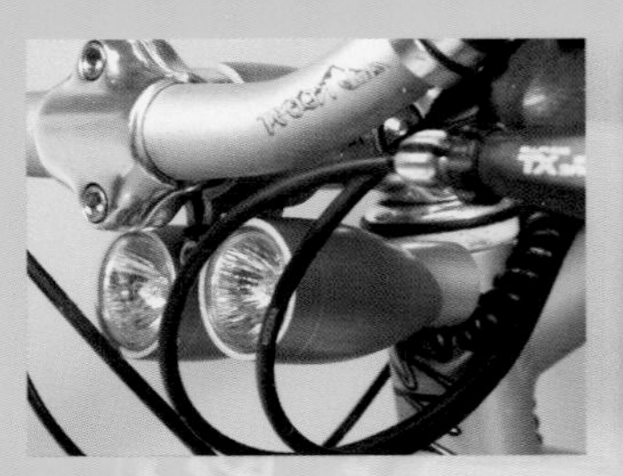

» bikerlicht «
» christian zwinger «

--- hochrobuste mountainbikebeleuchtung, die mit wenigen handgriffen in eine leistungsstarke handlampe umgewandelt werden kann ---

--- highly robust mountain bike lighting that can be converted to a powerful hand lamp with a few adjustments ---

» gangster «
» barbara keimel «

--- einfach montierbar auf verschiedenen helmmodellen an verschiedenen stellen -- befestigung durch saugnäpfe -- reflektorteil ist beweglich -- kabelaufwicklung unter dem akkufach, d.h. kabel hängt nicht durch ---

--- easily mounted at different places on various helmet models -- attached by means of suction cups -- reflector part is moveable -- cable rolled up under the built-in rechargeable battery, which means that the cable does not sag or trail ---

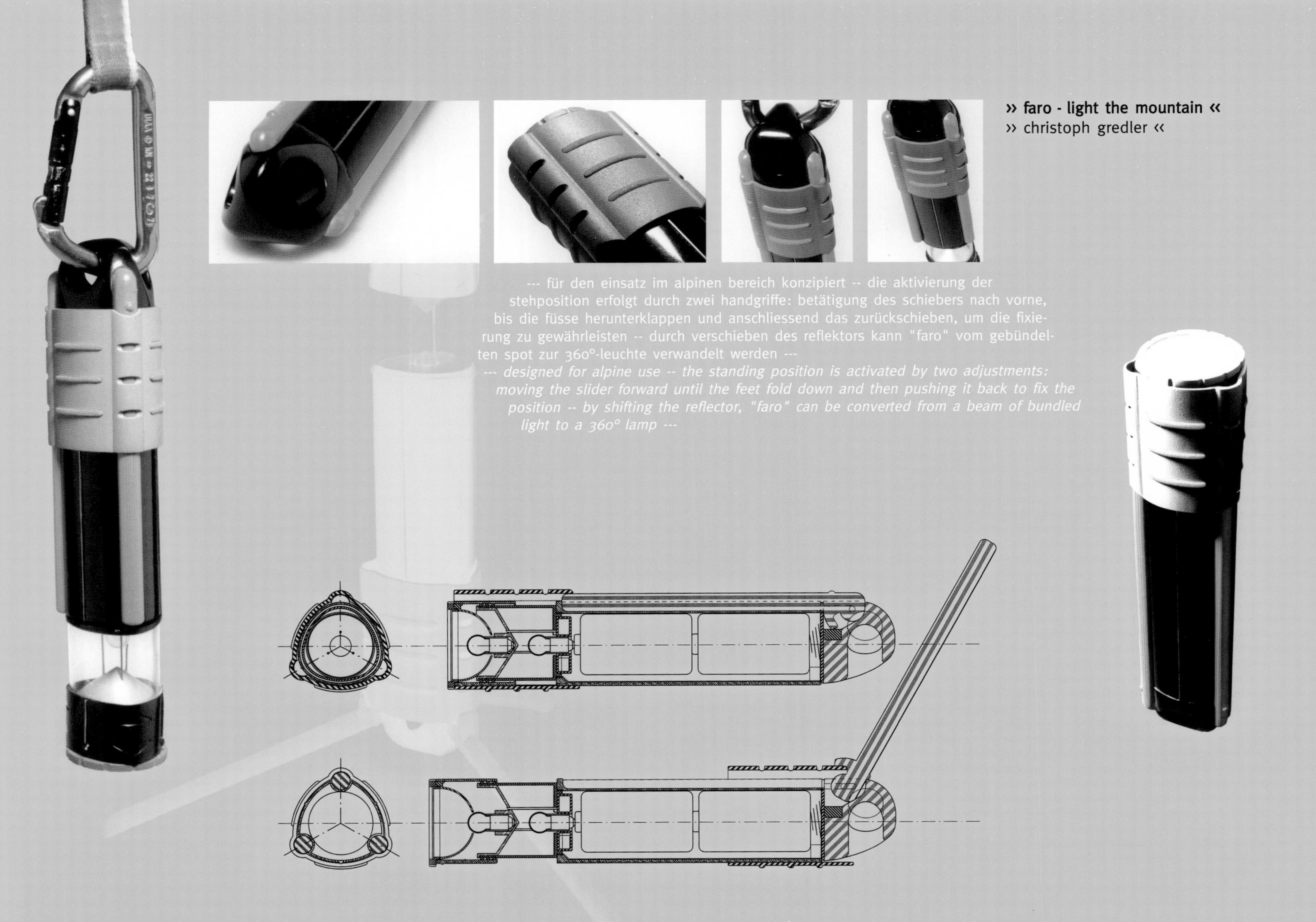

›› faro - light the mountain ‹‹
›› christoph gredler ‹‹

--- für den einsatz im alpinen bereich konzipiert -- die aktivierung der stehposition erfolgt durch zwei handgriffe: betätigung des schiebers nach vorne, bis die füsse herunterklappen und anschliessend das zurückschieben, um die fixierung zu gewährleisten -- durch verschieben des reflektors kann "faro" vom gebündelten spot zur 360°-leuchte verwandelt werden ---

--- designed for alpine use -- the standing position is activated by two adjustments: moving the slider forward until the feet fold down and then pushing it back to fix the position -- by shifting the reflector, "faro" can be converted from a beam of bundled light to a 360° lamp ---

1998 (5)

projekt: » mensch auf rollen «
4. semester / sommer 1998

betreuer: kurt h i l g a r t h (fancy form) - gastprofessor
georg w a g n e r - professor für engineering -
werner k l e i s s n e r - walter l a c h - modellbau

thema: der rahmen für diese aufgabenstellung ist bewusst sehr weit gesteckt: menschen, die sich mit ihrer muskelkraft auf rollen fortbewegen wollen. das kann vom arbeitseinsatz bis zum freizeitbereich gehen. zwei rahmenbedingungen gilt es einzuhalten: erstens darf der rollendurchmesser 100 mm nicht überschreiten und zweitens darf ausser der eigenen muskelkraft keine fremdenergie eingesetzt werden.

project: » people on rollers «
4th semester/spring 1998

advisors: kurt h i l g a r t h (fancy form) - guest professor -
georg w a g n e r - professor of engineering -
werner k l e i s s n e r - walter l a c h - model building -

subject: the scope for this task was purposely very broadly based: people who want to move around on rollers using muscle power. this can include use in work situations, as well as during leisure-time activities. there are two peripheral conditions that need to be observed: first of all, the diameter of the rollers may not exceed 100 mm and, secondly, the only source of energy may be the user's own muscle power.

mensch auf r

sign

projektarbeiten des fh-studienganges industrial design-graz
projects of the fh-degree course industrial design-graz

1998
(5)

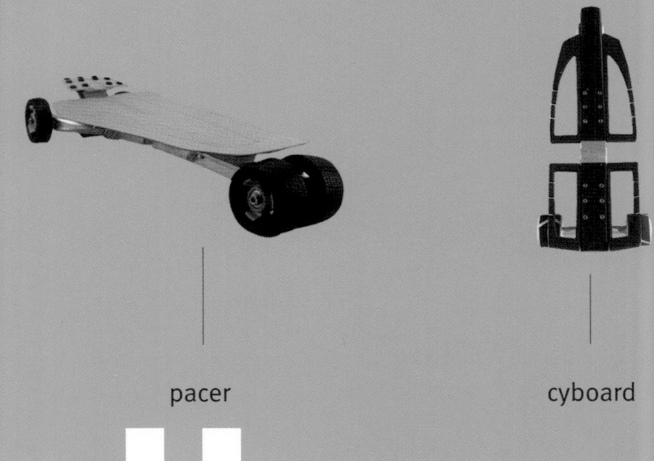

pacer

cyboard

dr. martens air wear

cityskater

smoothers

ollen

» junior skates «

» barbara keimel «

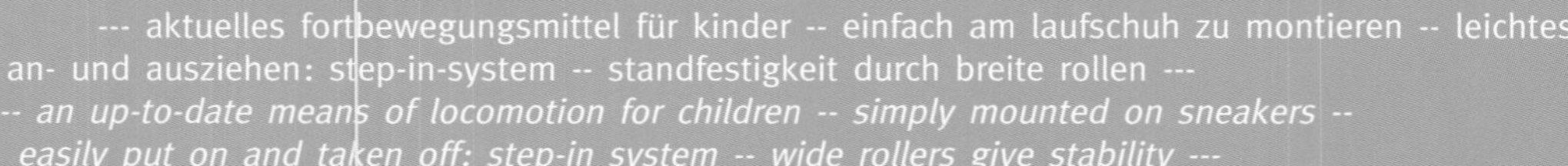

--- aktuelles fortbewegungsmittel für kinder -- einfach am laufschuh zu montieren -- leichtes an- und ausziehen: step-in-system -- standfestigkeit durch breite rollen ---

--- an up-to-date means of locomotion for children -- simply mounted on sneakers -- easily put on and taken off: step-in system -- wide rollers give stability ---

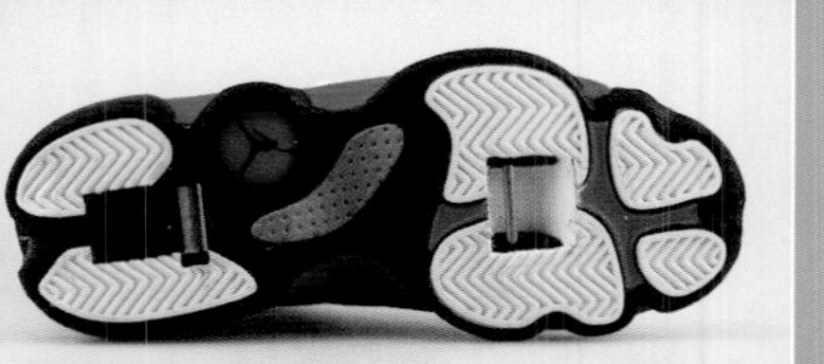

--- tiefer schwerpunkt -- verwendung mit normalen strassenschuhen -- schnelleinstieg durch vorklappbare "bindung" -- zielgruppe: - easy skater ---

--- low center of gravity -- use with normal street shoes -- "binding" that tilts forward makes quick "in and out" possible -- target group: - skaters looking for some easy skating ---

» cocroach «

» leonard natterer «

--- "pacer": selbstlenkende hinterachse -- dynamisch durch extrem niedrige schwerpunktslage -- formal an der zielgruppe der 18-25-jährigen orientiert ---

--- "pacer": self-steering rear axle -- extremely low center of gravity makes it dynamic -- designed for the target group of the 18-25 year-olds ---

›› pacer ‹‹

›› petrus gartler ‹‹

›› cityskater ‹‹

›› thomas binder ‹‹

--- dreirollenskaterschuh -- der bremsklotz ist dämpfung und verbindung von schale und achse in einem -- lenkung durch elastisch gelagerte hinterachse -- mit dem "cityskater" schnell durch das städtechaos! ---

--- three roll skater shoe -- the brake block combines both attenuation and the connection between shell and axle -- the flexibly placed rear axle takes care of the steering -- use the "city skater" to negotiate chaotic urban conditions quickly! ---

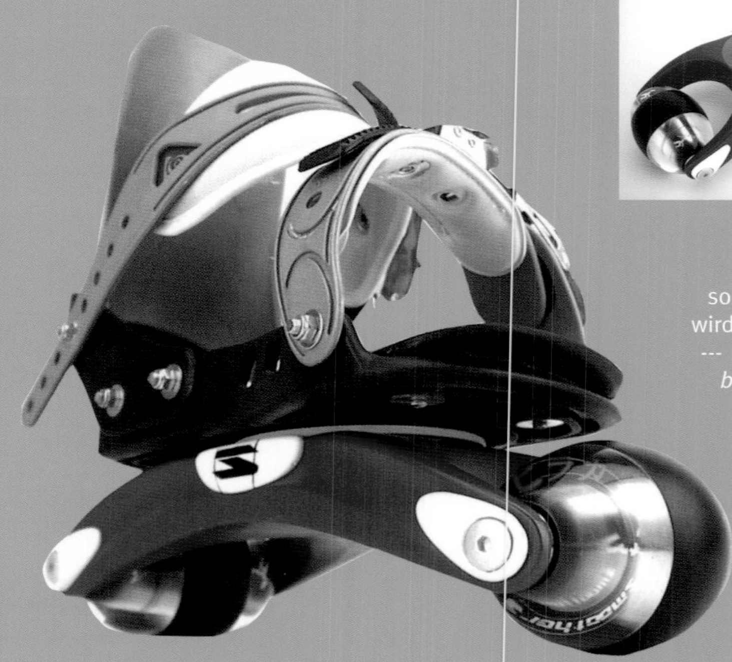

» smoothers «
» julian hönig «

--- grundidee: ein zweiteiliges system aus fahrgestell und snowboardbindung -- daraus ergibt sich mit dem gleichen fahrgestell eine wintervariante und eine sommervariante (bei der eine eigene bindung für strassenschuhe angeboten wird) ---

--- basic idea: a two-part system made up of a chassis and a snowboard binding -- this makes it possible to have both a winter model and a summer model using the same chassis (a separate binding is available for street shoes) ---

» cyboard «
» christian zwinger «

--- lenkgeometrie: "snaking extreme": - optimaler vortrieb beim fahren von schlangenlinien -- doppelrollen sorgen für schnelle kurvenfahrten durch satte bodenhaftung ---

--- steering geometry: "snaking extreme": - optimal propulsion when skating in curved lines -- double rollers guarantee excellent road adherence and make fast curves possible ---

» board «
» christian becker «

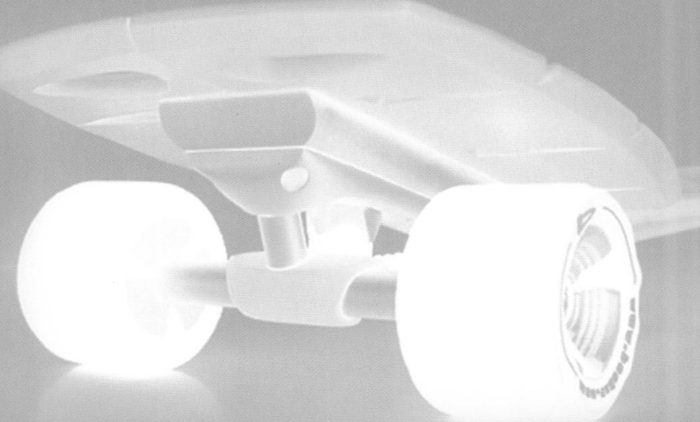

--- dreirad skateboard für innenstadt und schulweg -- durch gewichtsverlagerung selbstlenkende hinterachse -- auswechselbare, verschraubte trittflächen ---

--- three-wheel skateboard for downtown and the way to school -- self-steering rear axle through shifting of weight -- interchangeable, screwed on treads ---

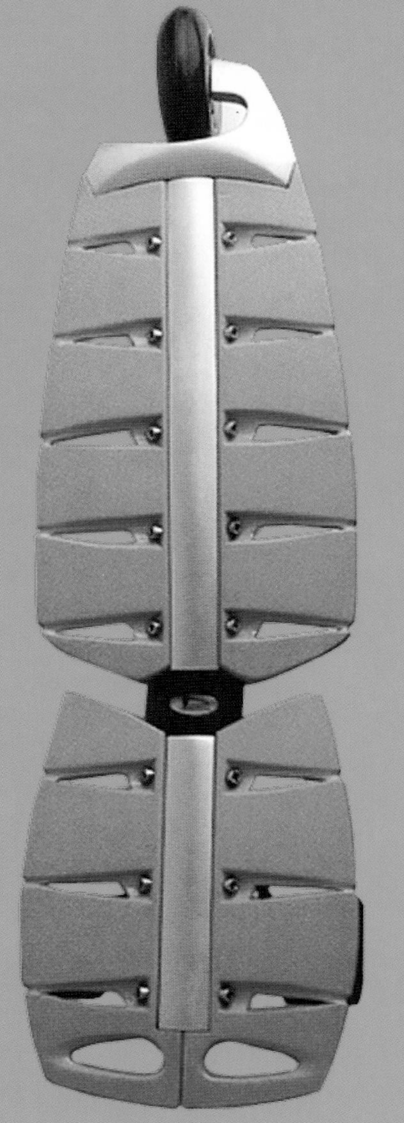

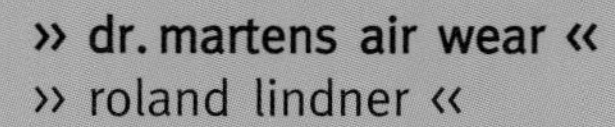

» dr. martens air wear «
» roland lindner «

--- zielgruppe: - träger von "dr. martens air wear boots" -- anforderungen: schnell an- und abnehmbar -- die beschaffenheit der sohle ermöglicht die verbindung mit dem system (...stahlseilbindung...) ---

--- target group: - "dr. martens air wear boots wearers" -- requirements: quick on and off -- the composition of the sole makes the connection with the system possible (...steel cable binding...) ---

projekt: » elektro-kleinfahrzeuge «
- in kooperation mit att/wachauer
6. semester/sommer 1998

betreuer: franz l e c h e r (mercedes benz advanced design) - gastprofessor
georg w a g n e r - professor für engineering -
werner k l e i s s n e r - walter l a c h - modellbau -

thema: für den zustelldienst im innerstädtischen bereich – unter einbeziehung von fussgängerzonen – soll ein umweltfreundliches leichtfahrzeug entwickelt werden. einsatzmöglichkeiten sind die zustellung von postpaketen, pizzas, blumen u.ä.. die breite soll 80 cm nicht überschreiten, die möglichkeit eines austauschbaren containers gegeben sein. wetterschutz gegen regen und schnee für fahrer/in ist voraussetzung.

project: *» small electrical vehicles «*
- in collaboration with att/wachauer
6th semester/spring 1998

advisors: *franz l e c h e r (mercedes benz advanced design) - guest professor*
georg w a g n e r - professor of engineering -
werner k l e i s s n e r - walter l a c h - model building -

subject: *development of an environmentally-friendly light vehicle for deliveries in the inner cities is planned, keeping in mind that today, many cities have pedestrian zones. deployment possibilities are deliveries of mail and packages, pizza, flowers, etc. the width should not exceed 80 cm, and there should be a possibility to use an exchangeable container. a definite requirement is protection against the elements for the driver.*

elektro-kleinfa

sign

projektarbeiten des fh-studienganges industrial design-graz
projects of the fh-degree course industrial design-graz

1998
(6)

elevant

gamma

hrzeuge

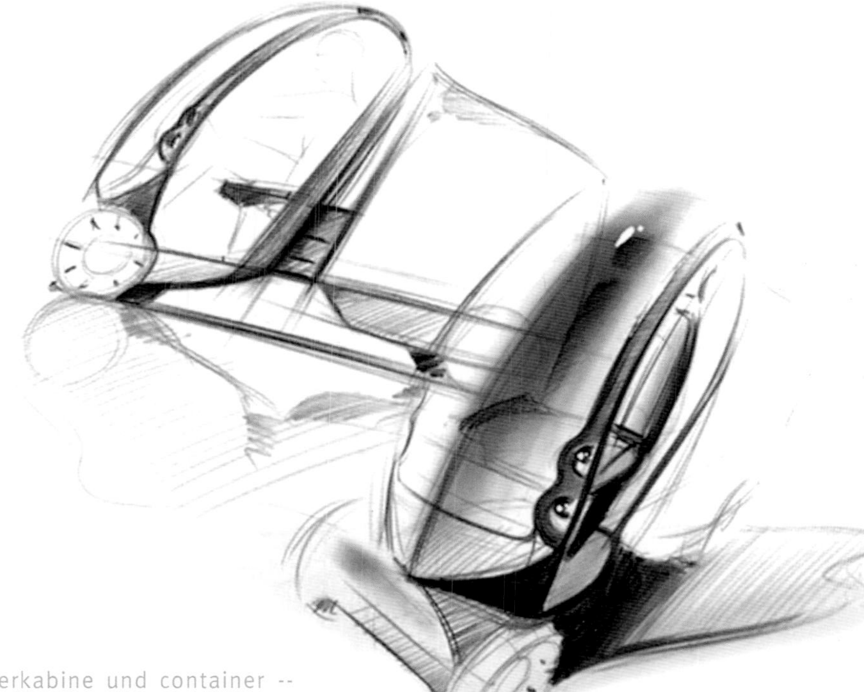

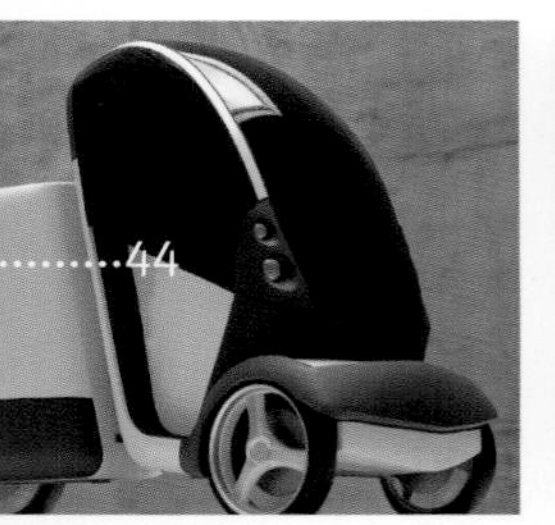

--- knicklenker "2-teilung": fahrerkabine und container --
türen: nach oben klappbar -- antriebseinheit vorne, gute zugänglichkeit bei akkuwechsel ---

*--- vehicle with articulated steering, "bi-partitioned": driver's compartment and container --
doors: hinged upwards -- propulsion unit in the front and easily accessible when
changing rechargeable battery ---*

--- vierrädriges leichtkraftfahrzeug -- länge 250 cm, breite 80 cm (türdurchfahrt), höhe 195 cm -- geschlossener aufbau (wetterschutz) -- transportfunktion (zustelldienste wie post, pizza, blumen usw.) -- formale unterscheidung gegenüber pkw -- elektroantrieb ---

--- light four-wheel vehicle -- 250 cm long, 80 cm wide (door width), 195 cm high -- closed superstructure (weatherproof) -- used for transport (delivery services such as mail, pizza, flowers, etc.) -- different design than regular cars -- electrically powered ---

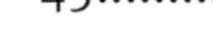

-- alias visualisierung niki pelzl --

» elevant «

» sylvia feichtinger, tanja langgner, niki pelzl, david stelzer, michael tropper «

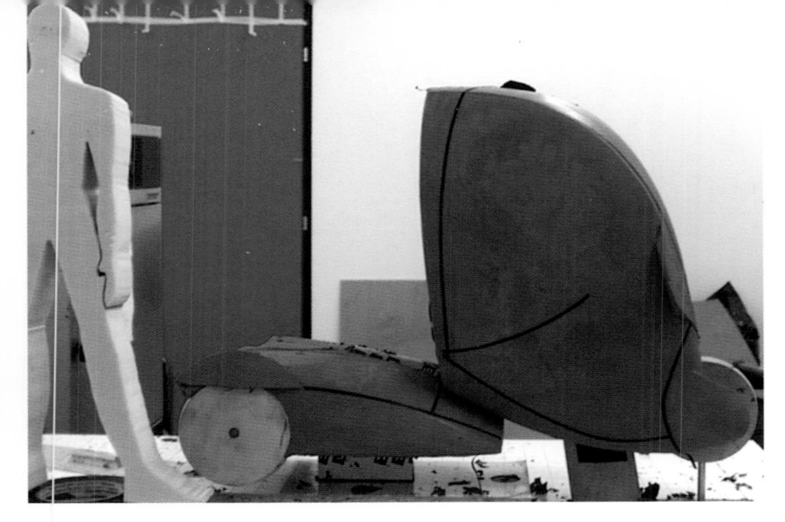

-- renderings --

-- clay modeling --

›› gamma ‹‹
›› johannes scherr, silke spath, klaudia takats, christian winkler ‹‹

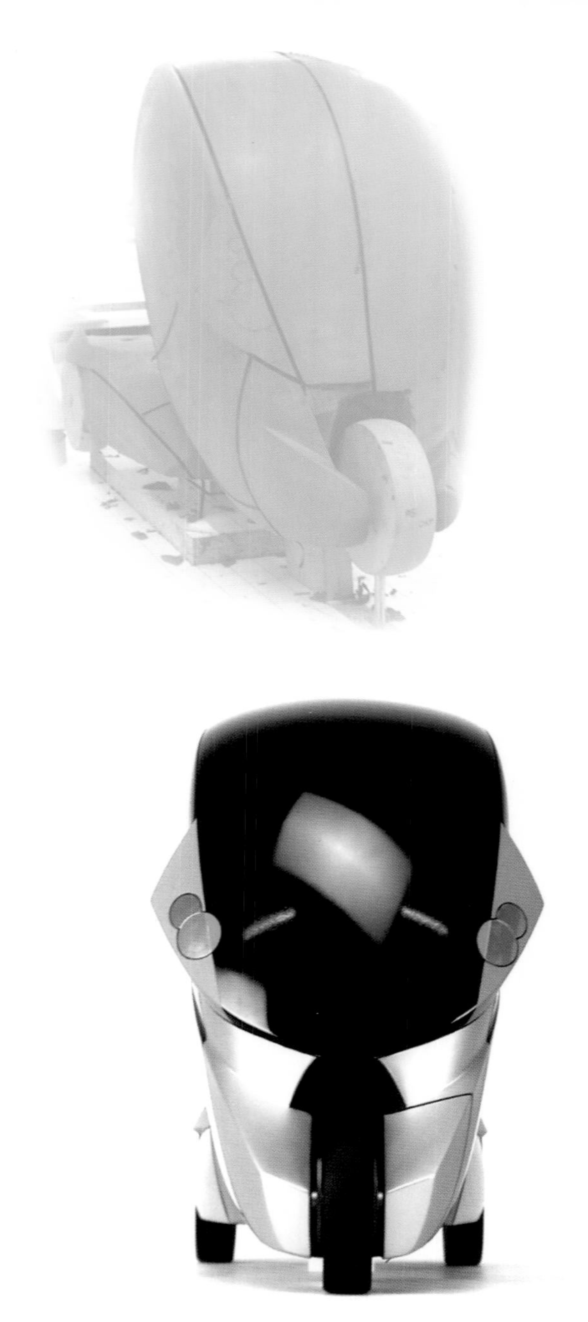

--- prinzip eines dreirädrigen rollers -- das fahrzeug besteht aus drei teilen: - die heckplattform mit der elektrischen antriebseinheit, - die austauschbaren container und - die fahrerkabine -- schwenkbar um eine horizontalachse: bei kurvenfahrt in fahrdynamische schräglage -- dynamische vorderradaufhängung durch integralkonzept mit fahrerkabine ---

--- uses the principle of a three-wheel scooter -- the vehicle consists of three parts: - the rear platform with the electrical propulsion unit, - the interchangeable container and - driver's compartment -- can be swiveled around a horizontal axle: dynamic handling in curves in an inclined position -- front wheel suspension is dynamic because of the integral approach in connection with the driver's compartment ---

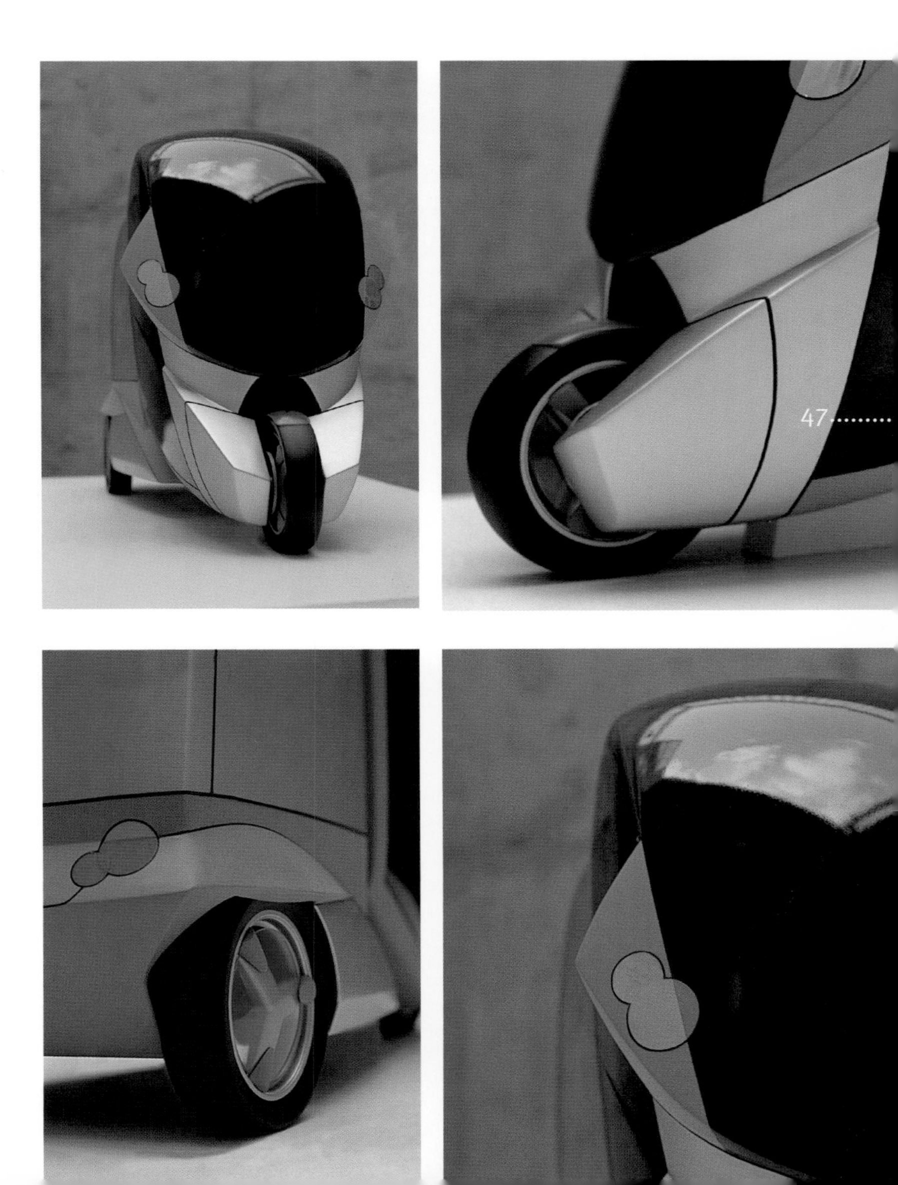

thema: der sponsor sucht innovative designideen für geschenke an besondere gäste bzw. jubilare, voraussetzung ist der werkstoff stahl.

wettbewerb: eine hochkarätige, international besetzte jury unter leitung von dirk schmauser (porsche design) vergibt bei 76 einreichungen drei preise, mehrere anerkennungen und ankäufe. der grazer studiengang kann dabei den 2. und 3. platz verbuchen (beate dörflinger und albert ebenbichler), einen anerkennungspreis (joe lintl) und zwei ankäufe (gerald krenn und marc ischepp). die "geheimnisvolle dose" von beate dörflinger wird als einziges projekt in einer limitierten serie realisiert.

anerkennungspreis, (lir

präsente in st

subject: *the sponsor of the competition is looking for innovative design ideas for gifts destined for special guests of the company and/or celebrations for long-time employees. the only prerequisite is the use of steel as the material of the object.*

competi-tion: *a top-class, international jury, led by dirk schmauser (porsche design), gave out three prizes and several honorable mentions to the 76 entrants. several entries were also purchased outright. the graz class was successful in placing second and third (beate dörflinger and albert ebenbichler), as well as an honorable mention (joe lintl) and two outright purchases (gerald krenn and marc ischepp). the "mysterious box" by beate dörflinger is the only project to be immediately realized; a limited series was made.*

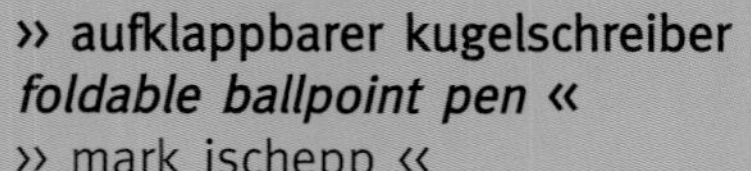

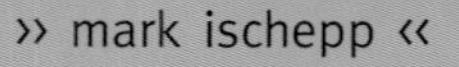

›› aufklappbarer kugelschreiber
foldable ballpoint pen «
›› mark ischepp ‹‹

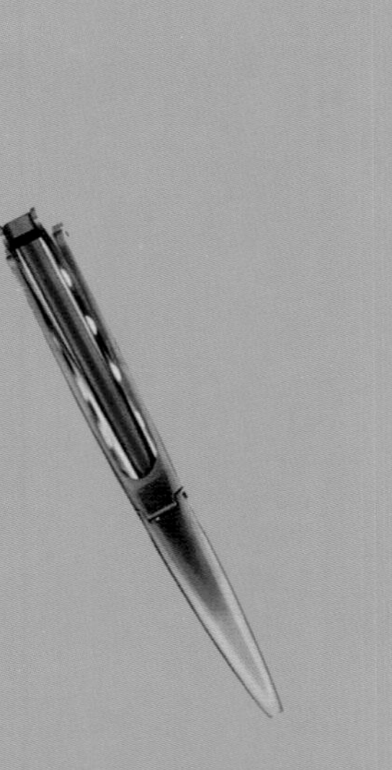

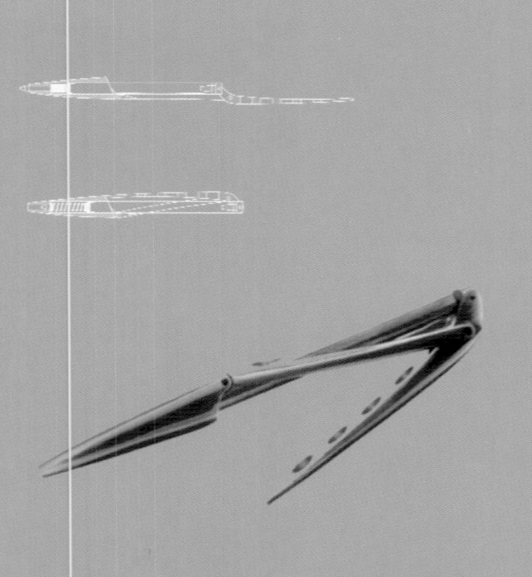

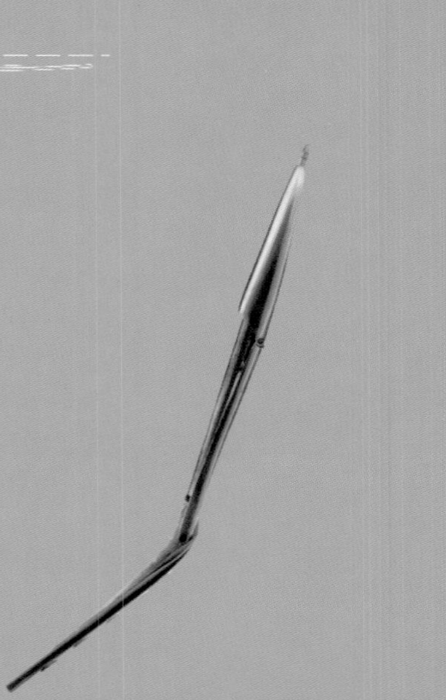

ankauf, *purchase* (krenn)

1. preis und realisierung, (beate dörflinger)
1st prize and realization

ankauf, *purchase* (ischepp)

1998/99 (7)

projekt: » präsente in stahl « - ideenwettbewerb für österreichische design-studenten, ausgeschrieben von der voest alpine stahl ag
3. semester/winter 1998/99,

project: » gifts in steel « - competition of ideas for austrian design students, sponsored by the voest alpine stahl ag
3rd semester/winter 1998/99

betreuer: gerhard h e u f l e r - gerald k i s k a - professoren für design -
georg w a g n e r - professor für engineering -
werner k l e i s s n e r - walter l a c h - modellbau -

advisors: gerhard h e u f l e r - gerald k i s k a - professors of design -
georg w a g n e r - professor of engineering -
werner k l e i s s n e r - walter l a c h - model building -

...hinstellen *put down*

...drehen *spin*

...öffnen *open*

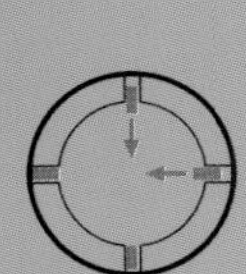

--- jury: faszinierend an diesem entwurf ist der verschlussmechanismus -- entriegelt werden kann der deckel nur, wenn der benutzer die gesetze der massenträgheit bzw. zentrifugalkraft berücksichtigt ---

--- jury: the closing mechanism was the fascinating thing in this design -- the lid can only be opened if the user takes the laws of inertia and/or centrifugal force into account ---

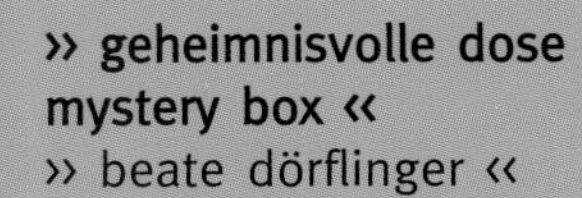

» geheimnisvolle dose
mystery box «
» beate dörflinger «

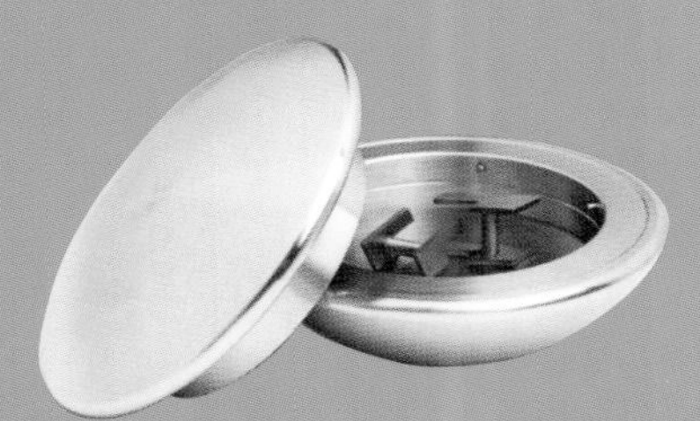

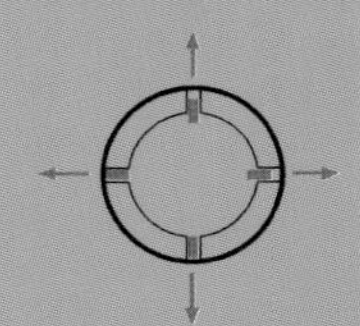

...neigen *tilt*

...und *and*

...wieder verschlossen *closed again*

wander navigator

1998/99 (8)

projekt: » wandernavigator «
- in koop. mit dem fh-studiengang "industrielle elektronik"
3. semester/winter 1998/99

project: » hiking navigator «
- in coop. with the fh degree course "industrial electronics"
3rd semester/fall 1998/99

betreuer: gerhard h e u f l e r - gerald k i s k a - professoren für design -
georg w a g n e r - professor für engineering -
werner k l e i s s n e r - walter l a c h - modellbau -

advisors: gerhard h e u f l e r - gerald k i s k a - professors of design -
georg w a g n e r - professor of engineering -
werner k l e i s s n e r - walter l a c h - model building -

thema: die wanderkarte aus papier hat ausgedient! der wandernavigator ist eine innovative kombination aus digitalisierter landkarte, gps, elektron. kompass und mikrodisplay (= grosses virtuelles bild!). ausgezeichnet mit dem "internationalen motorola design award 1999".

subject: the days of the hiking map made of paper are over! the hiking navigator is an innovative combination of a digital map, gps, electronic compass and a micro display (= large virtual image!). this device was honored with the "international motorola design award 1999".

» leader «
» peter kalsberger «

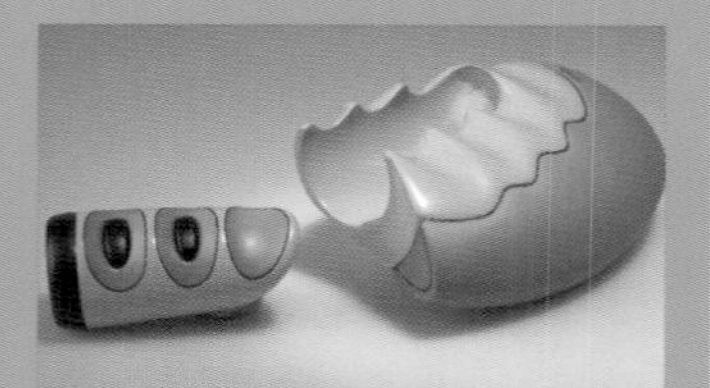

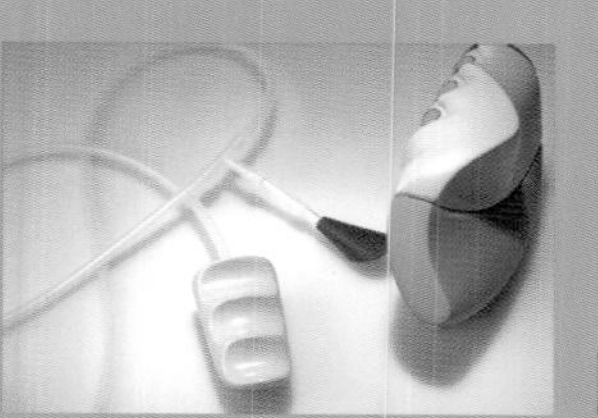

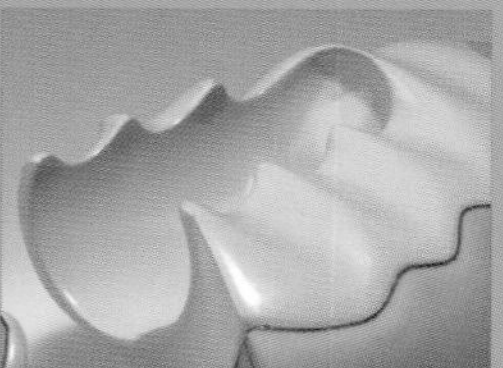

--- "leader" for freaky experience -- zweikomponentensystem: trennung des okulars von der elektronik -- verwendung: - kompakt für einfache touren, - separat für komplizierteres gelände ---

--- a "leader" for super experiences -- two component system: ocular device is separate from the electronics -- use: - compact for simple hikes, - different for complicated terrain ---

» okular «
» daniel dockner «

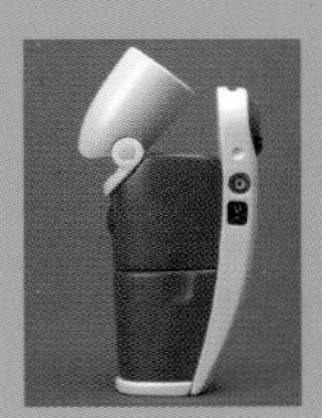

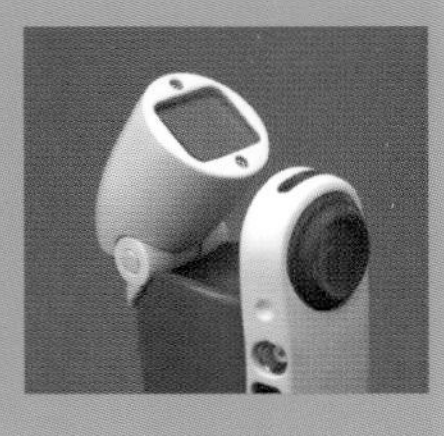

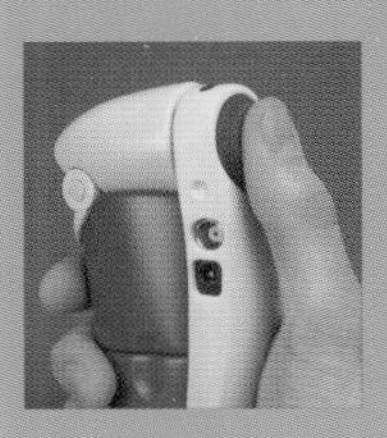

--- gps-empfänger mit integrierter digitaler landkarte für wissenschaft und forschung -- eine voice- und memofunktion zur lückenlosen erfassung aller details ---
--- gps receiver with integrated digital map for scientific and research purposes -- a voice and memo function for complete record of all details ---

--- speziell für - touren, - snowboarder und - mountainbiker entwickelt -- bedienelement hat joystick-funktion ---
--- especially for - hikes, -- developed for - snowboarders and - - mountain bikers -- user control functions as a joystick ---

» country guide «
» joe lintl «

» gps - navigator «
» mark ischepp «

--- "gps-navigator" für den stadttouristen -- einhand-bedienung mittels drehrad ---
--- "gps navigator" for the urban tourist -- one-hand operation by means of a control dial ---

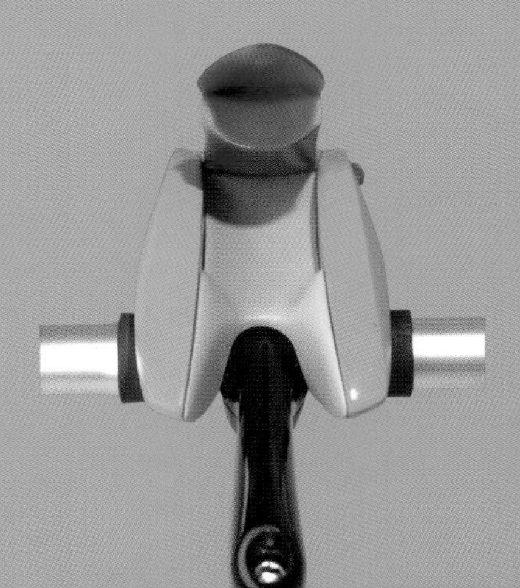

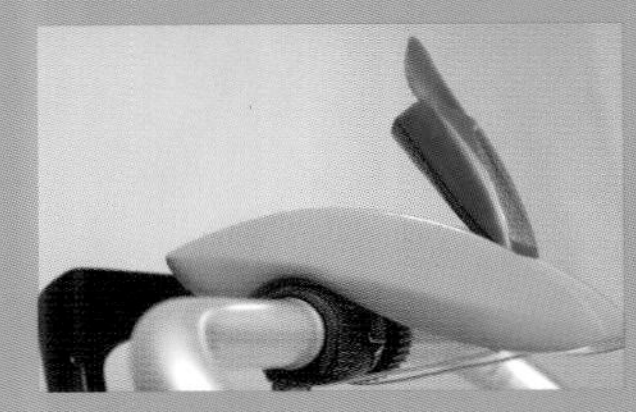

» gps für racing bikers «
» beate dörflinger «

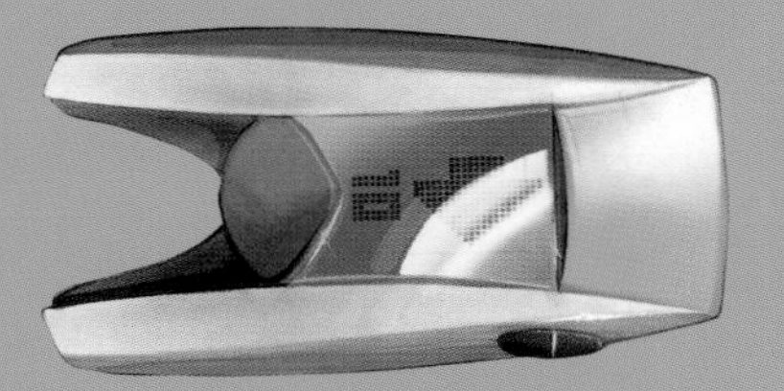

--- elektronische landkarte kombiniert mit herzfrequenz-, pulsmessung und tachometer sowie richtungsanweisungen zur navigation ---
--- electronic map combined with a measuring device for heart rate, pulse, and speedometer, as well as directional instructions for navigation ---

projekt: » traditionelles wohnen + neue medien «
7. semester/winter 1998/99

betreuer: yasushi k u s u m e + jan-eric b a a r s (philips design) - gastprofessoren -
georg w a g n e r - professor für engineering -
werner k l e i s s n e r - walter l a c h - modellbau -

thema: die meisten menschen bevorzugen auch heute noch traditionelle wohnformen. die neuen medien aber schaffen neue, unkonventionelle möglichkeiten im audio-visuellen bereich. wie lassen sich diese scheinbaren gegensätze auf einen gemeinsamen nenner bringen? wie reagieren unterschiedliche zielgruppen auf diese herausforderung? und welche rolle spielen dabei die verschiedensten umgebungsbezüge – von der küche über das wohnzimmer bis zum bad?

project: » traditional living and new media «
7th semester/fall 1998/99

advisors: yasushi k u s u m e + jan-eric b a a r s (philips design) - guest professors -
georg w a g n e r - professor of engineering -
werner k l e i s s n e r - walter l a c h - model building -

subject: most people still prefer living in a traditional environment, but the new media are creating new and unconventional audio-visual possibilities. how can we bring these apparent contradictions closer so that they have one common denominator? how do different target groups react to this challenge? and what kind of role does our relationship to different parts of our surroundings play – from the kitchen to the living room and to the bathroom?

frühstücksradio

sign

projektarbeiten des fh-studienganges industrial design-graz
projects of the fh-degree course industrial design-graz

1998/99
(9)

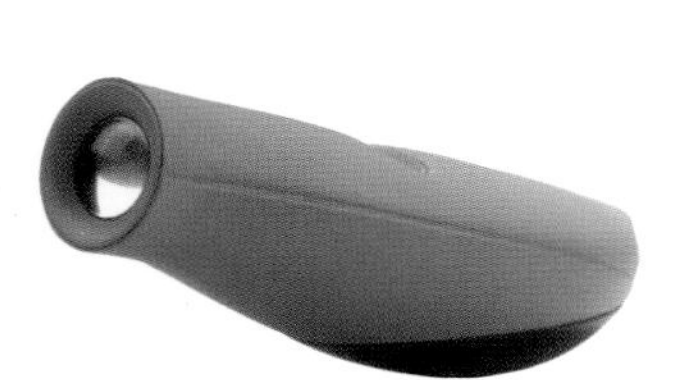

interface-vision 2

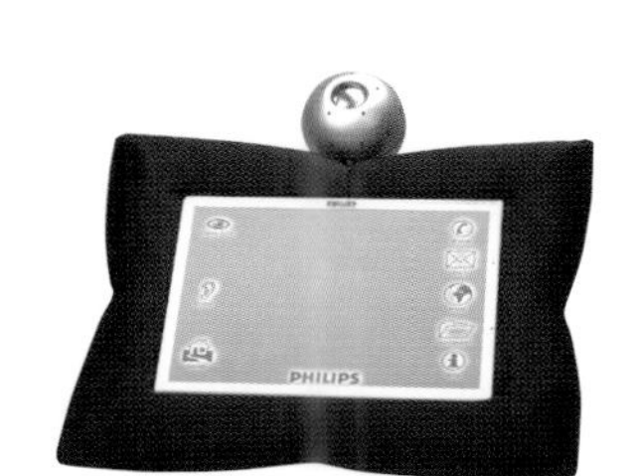

pillow

tv-book

blow it up

swoof

medien

» tv-book «
» tanja langgner «

--- aufklappbares "tv-book" durch flexiblen screen mit ladestation -- schutz beim transport durch buchhülle ---
--- fold-out "tv book", made possible by a flexible screen with a charging station -- the book cover protects the device during transport ---

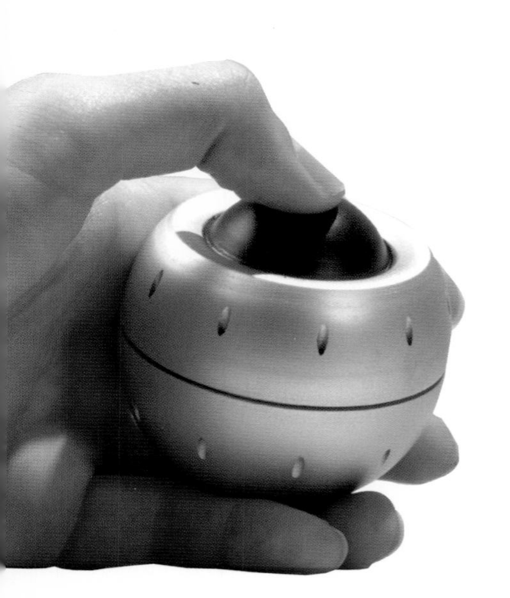

» pillow «
» gernot puschner «

--- multifunktionale fernbedienung, mit der man unterschiedliche geräte bedient -- das gerät ist mit seiner kleinen abnehmbaren kamera auch ein video-phone ---
--- multifunctional remote control that operates various devices -- the device has its own small detachable camera and a video phone ---

» interface - vision 1 «
» michael tropper «

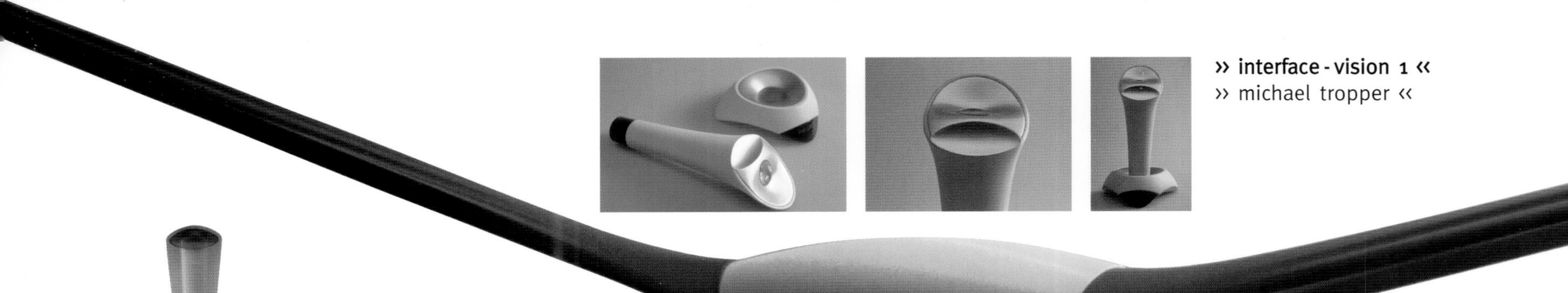

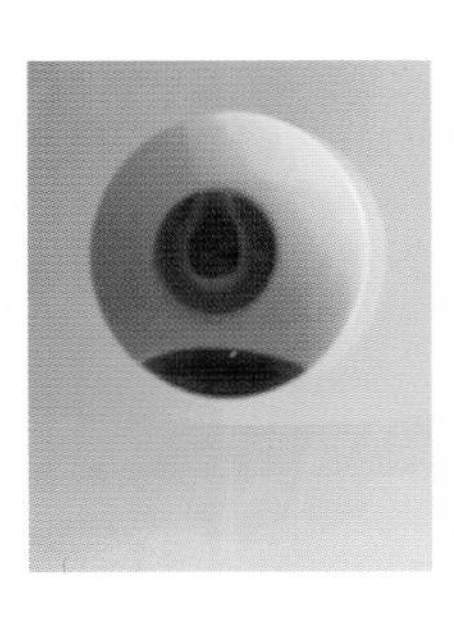

--- schlafzimmer-küche-badezimmer-wohnzimmer: in jedem dieser räume befindet sich eine speziell abgestimmte fernbedienung zur steuerung aller unterhaltungsgeräte -- fernbedienung: der schlüssel zur bedienung ist ein navigationsbild, welches vom benutzer an eine beliebige fläche projiziert wird -- über ein touch-pad wird ein cursor auf dieser projektion bewegt und die verschiedenen menü-funktionen können aktiviert werden -- textilien, verspannte seile und stangen wurden zum bestimmenden gestaltungselement für die raumspezifischen klangkörper ---

--- bedroom-kitchen-bath-living room: each one of these rooms has its own specially adjusted remote control for all home entertainment electronics -- remote control: for operation, a navigational image is projected by the user onto any flat surface -- a cursor operated on a touch pad moves over this projected image and activates the various menu functions -- textiles, fixed ropes and rods are the defining design elements for the body of sound that is specific to each room ---

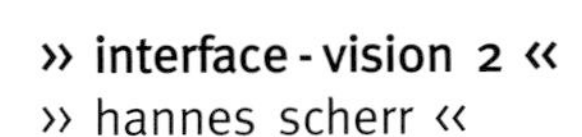

» interface - vision 2 «
» hannes scherr «

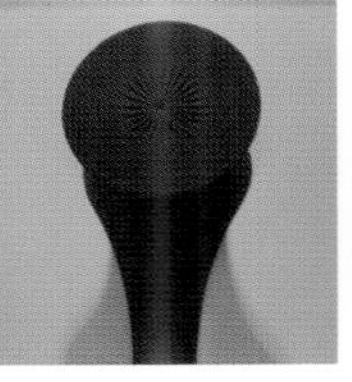

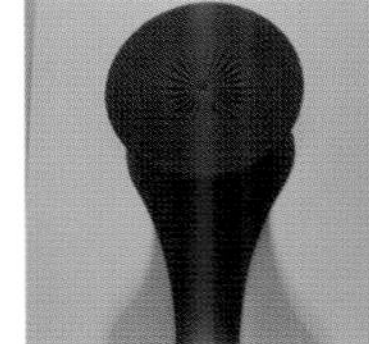

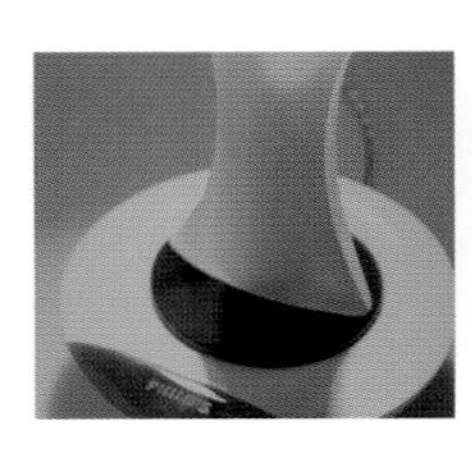

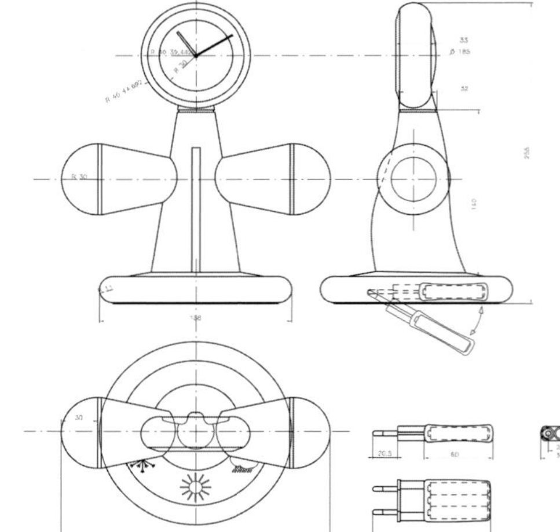

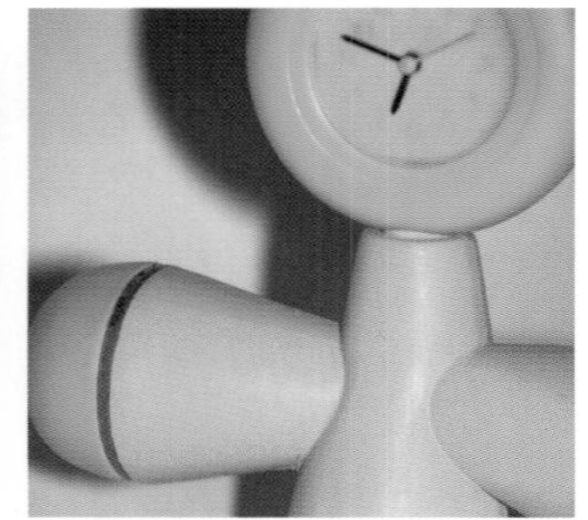

›› frühstücksradio ‹‹

›› oskar kalamidas ‹‹

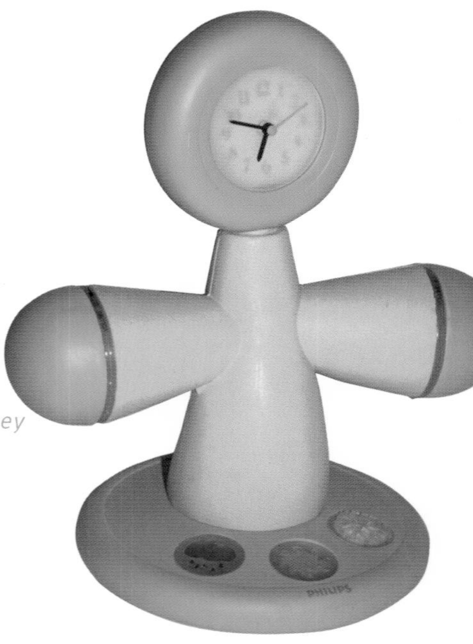

--- das "frühstücksradio" mit uhr und wettervorhersage -- lautstärkenregelung durch drehbaren seitlichen lautsprecher -- bei der gestaltung der bedienelemente wurde zusätzlich darauf geachtet, dass sie auch von menschen mit körperlichen einschränkungen verwendet werden können (tastbares zifferblatt der uhr, taktile anzeige der wettervorhersage) ---

--- early "morning radio" with time and weather information -- the loudspeakers on the sides, which can be swiveled, control the volume -- the design of the user controls is such that they can also be used by physically challenged persons (clock-face is a touch keypad, tactile display of the weather forecast) ---

›› blow it up ‹‹

›› sylvia feichtinger ‹‹

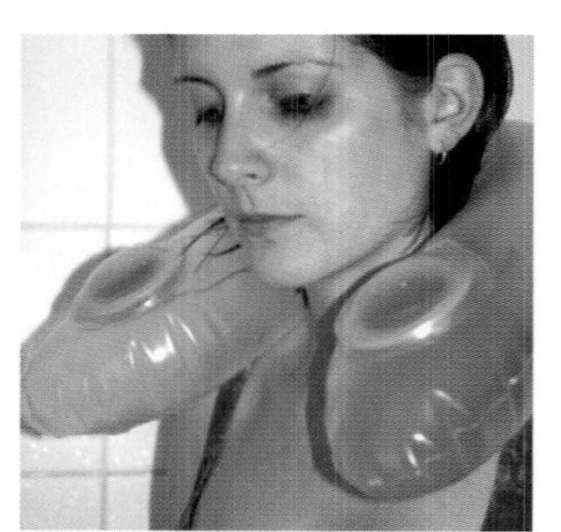

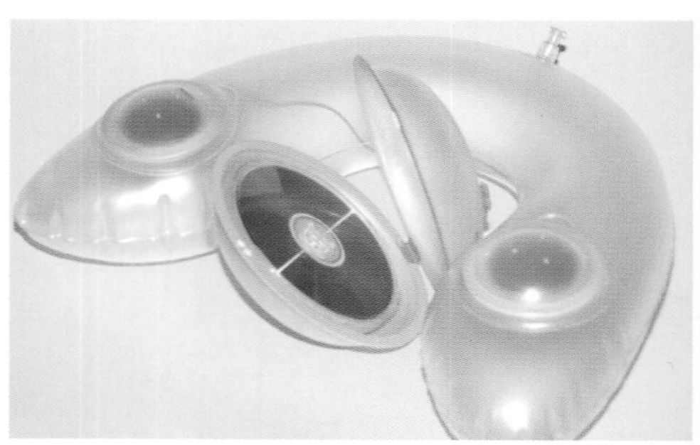

--- wasserfester cd-player und lautsprecher-system -- material: - pvc geschweisst -- an allen produkten befinden sich saugnäpfe, die an glatten flächen montierbar sind ---

--- waterproof cd player and loudspeaker system -- material: - welded pvc -- all products have suction cups that can be mounted on smooth surfaces ---

» swoof «
» niki pelzl «

--- fusion eines bequemen sitzmöbels mit einem high-end-speakersystem -- die lautsprecher können entlang des äusseren ringes im 120-grad-raster positioniert werden ---
--- combination of a comfortable chair or couch with a high-end speaker system -- the loudspeakers can be positioned around the outer perimeter in a 120 degree pattern ---

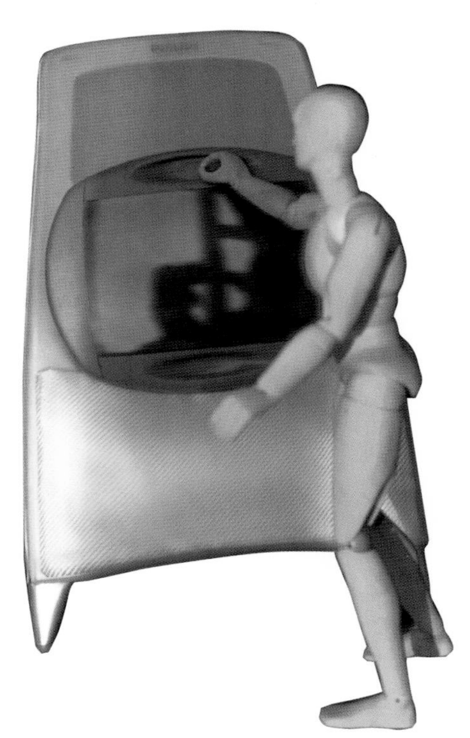

» interface panel «
» harald auer «

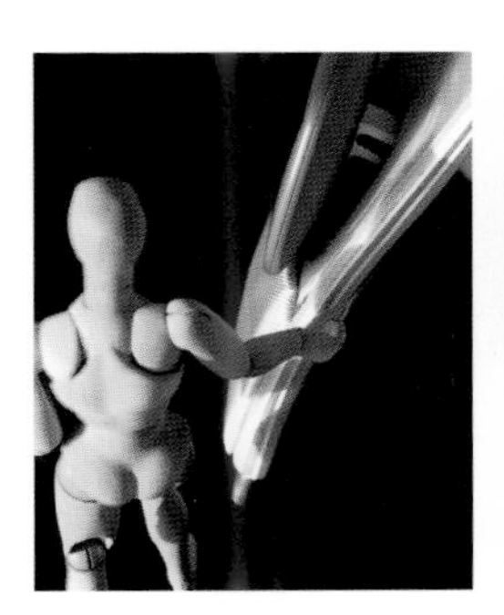
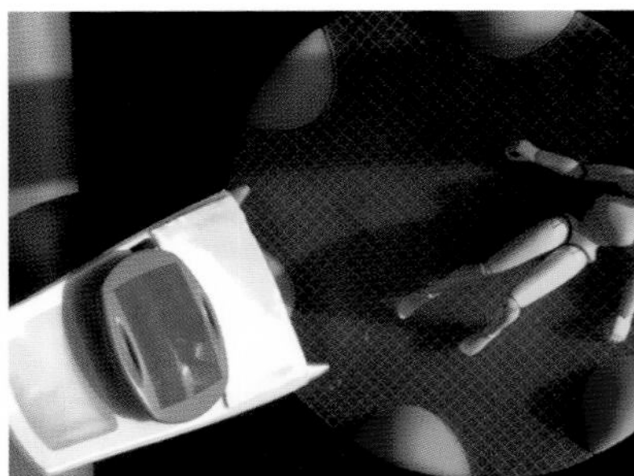

--- das mobile "interface-panel" wird am griff aus der flachen ruheposition an der wand auf den runden induktiven teppich (recreation area) gezogen und beliebig darauf kabellos positioniert ---
--- the mobile "interface panel" is pulled from its resting position on the wall by its handle and placed in any position on the round inductive rug (recreation area) ---

design-szenari

1999 (10)

projekt: » design-szenarien «
4. semester/sommer 1999,

betreuer: mathias p e s c h k e + james s k o n e - gastprofessoren
werner k l e i s s n e r - walter l a c h - modellbau

thema: wähle ein szenario, das du gut kennst! analysiere den ablauf ("walkthrough" methode), nenne die kriterien, nach denen etwas verändert werden soll (mission statement). nenne und entwickle lösungen (radikale ideen, utopien, möglichkeiten des verzichtes, dienstleistungen, produkte etc.). reflektiere über das projekt, führe tagebuch über die höhen und tiefen des kreativitätsprozesses. stelle diesen prozess als kurve dar, um sie mit späteren projektabläufen vergleichen zu können.

project: *» design scenarios «*
4th semester/spring 1999,

advisors: *mathias p e s c h k e + james s k o n e - guest professors*
werner k l e i s s n e r - walter l a c h - model building

subject: *choose some well-known scenario. analyze its sequence according to the "walkthrough" method, list the criteria according to which something needs to be changed (mission statement) develop and list solutions (radical ideas, utopias, possibilities of doing without something, services, products, etc.). think about the project and keep a diary about the highs and lows of the creativity process. make a graphic representation of this process in the form of a curve so that you can compare it with the actual project process later on.*

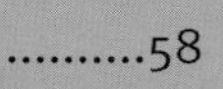

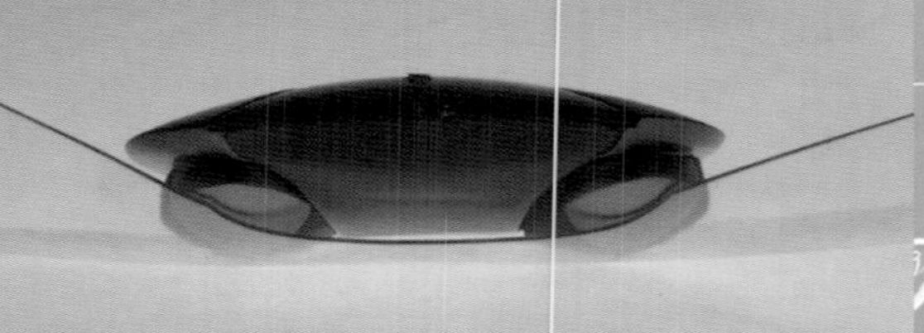

» suono «
» beate dörflinger «

--- lebensfreude ausdrücken, spontan tanzen, die musik fühlen -- töne werden über sensoren auf die haut übertragen --
die übertragung findet drahtlos statt, die sensoren sind schmuckelemente, die man entweder direkt am körper trägt, oder die in die kleidung sichtbar integriert werden -- elektronik wird somit zu einem modefeature ---

--- expressing zest for life, dancing spontaneously, feeling the music -- sounds are transmitted to the skin surface by sensors -- the transmission is wireless, the sensors are decorative elements that are worn directly on the body or are visibly integrated into the clothing -- thus electronics becomes a fashion accessory ---

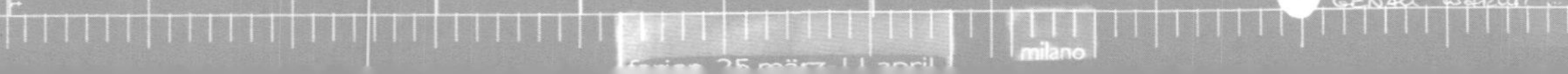

en

» scan vas «
» marc ischepp «

--- ein erlebnis, ein bild "festhalten" -- das bild, das man am flachen bildschirm sieht, hält man mit dem auslöser fest und hängt es zu hause an die wand, ohne es verändern zu können -- nach sieben tagen erlischt das bild; ... zeit, sich ein neues zu suchen ---

--- "preserving" an experience, an image -- you preserve the picture that you see on the flat screen using the release mechanism, and you can put it on your wall at home without being able to change it -- the image fades away after seven days; ... time to look for a new one ---

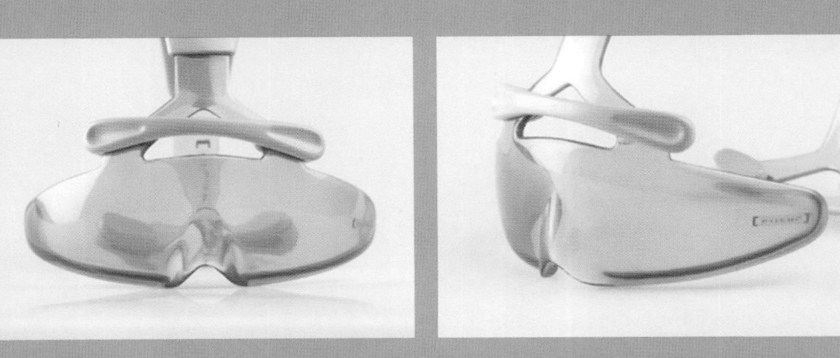

» [filtar] «
» peter kalsberger «

--- lifestyle, rave, party ... -- dj's geben der musik ihre persönliche note durch veränderung und manipulation -- die optische wahrnehmung wird (analog zur musik) mittels elektronischer impulse, drahtlos, über eine "brille" durchgeführt ---

--- lifestyle, rave, party ... -- dj's give the music their own personal touch by manipulating and changing it -- the visual perception is carried out wirelessly (analogously to the music) by means of electronic impulses through a pair of "eyeglasses" ---

FH JOANNEUM
FACHHOCHSCHULSTUDIENGÄNGE
DER TECHNIKUM JOANNEUM GMBH

1999 (11)

projekt: » pkw-anhänger «
6. semester/sommer 1999

betreuer: gerhard h e u f l e r - gerald k i s k a - professoren für design -
georg w a g n e r - professor für engineering -
werner k l e i s s n e r - walter l a c h - modellbau –

thema: pkw-anhänger sind technisch, optisch und vom gebrauchswert her auf dem entwicklungsstand der 60er jahre stehen geblieben. sie sind kein fokuspunkt der autoindustrie. dabei könnten pkw-anhänger den trend zum kompakten auto unterstützen: anstatt ständig mit viel zu grossem volumina durch die gegend zu fahren, können anhänger bei bedarf zusätzlichen stauraum bieten und in ihrer ausführung speziell auf die bedürfnisse ihrer benutzer hin konzipiert sein.

project: *» passenger car trailers" «*
6th semster/spring 1999

advisors: *gerhard h e u f l e r - gerald k i s k a - professors of design -*
georg w a g n e r - professor of engineering -
werner k l e i s s n e r - walter l a c h - model building -

subject: *as far as technology, appearance and practical use are concerned passenger car trailers have stayed at the same stage of development as they were in the 1960's. they are not a focal point of the automotive industry. however, passenger car trailers could support the trend toward compact cars. instead of constantly driving around with a large-capacity vehicle, trailers would offer additional space when needed. they would be designed with the needs of their users in mind.*

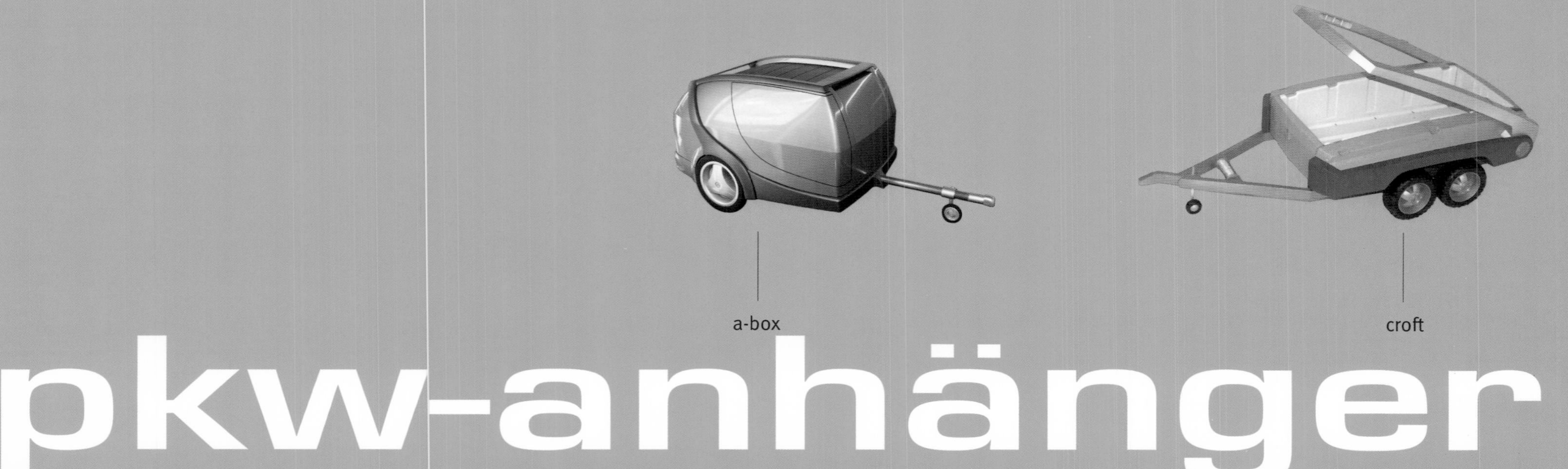

pkw-anhänger

sign

p r o j e k t a r b e i t e n d e s f h - s t u d i e n g a n g e s i n d u s t r i a l d e s i g n - g r a z
p r o j e c t s o f t h e f h - d e g r e e c o u r s e i n d u s t r i a l d e s i g n - g r a z

1999
(11)

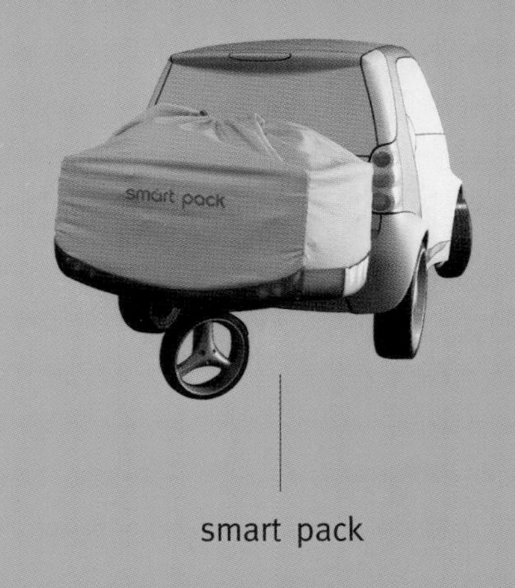

smart pack

audi st sport trailer

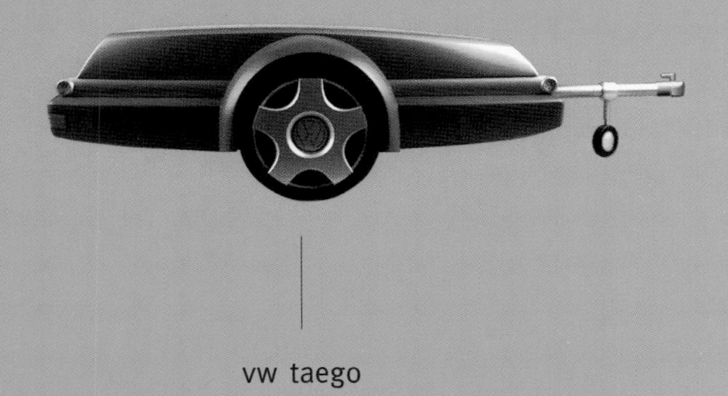

vw taego

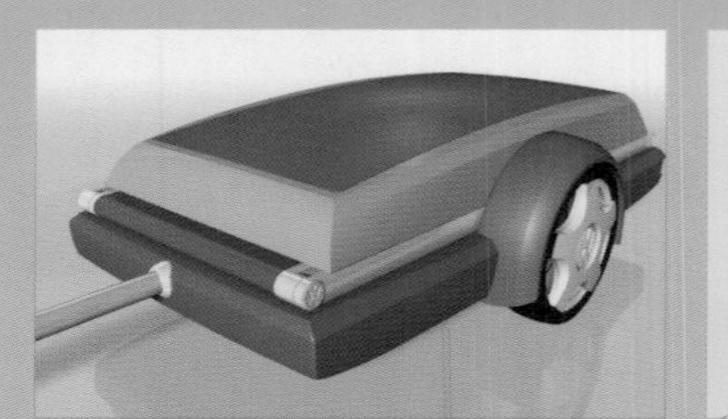

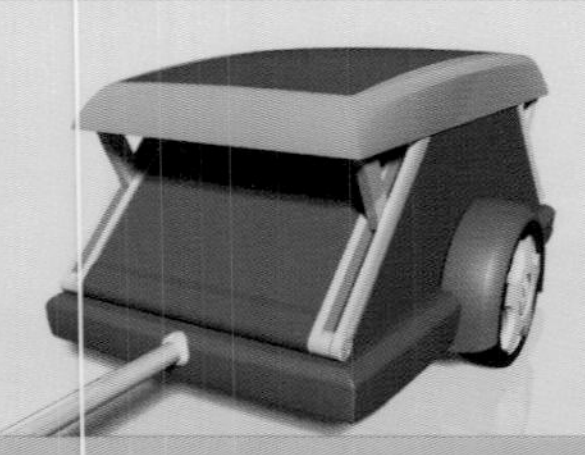

» vw taego «

» rudolf steiner, karen rosenkranz, valerie trauttmansdorff «

--- für vw-kompaktwagenfahrer, insbesondere junge, unternehmungslustige und flexible menschen -- durch eine hubmechanik lässt sich der anhänger mit nur geringem kraftaufwand aufklappen -- geometrisch exakte linienführung, die sich vom lamellendach bis unter den ladeboden zieht ---

--- for drivers of vw compacts, in particular young, adventurous and flexible people -- a lifting mechanism makes it possible to open the trailer with a minimum of effort -- a geometrically exact design from the lamellar roof to beneath the floor of the loading surface ---

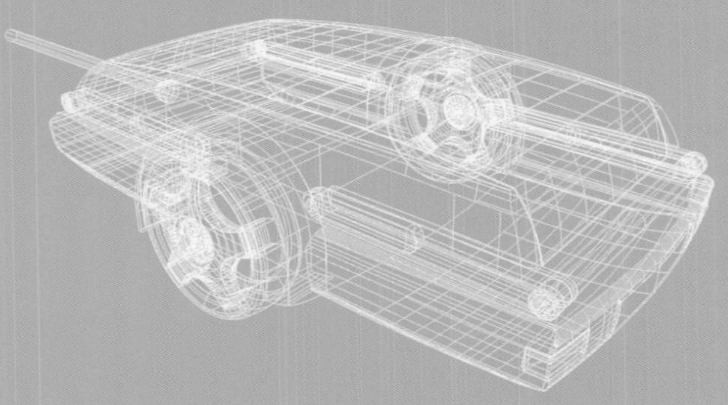

» a-box «

» barbara keimel, karin ruthardt, martin wöls «

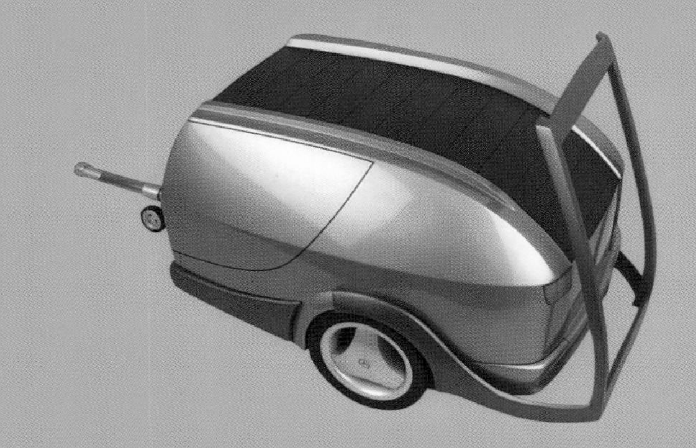

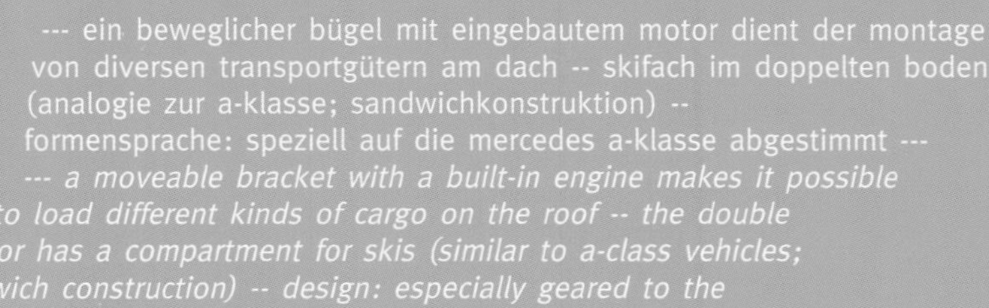

--- ein beweglicher bügel mit eingebautem motor dient der montage von diversen transportgütern am dach -- skifach im doppelten boden (analogie zur a-klasse; sandwichkonstruktion) -- formensprache: speziell auf die mercedes a-klasse abgestimmt ---

--- a moveable bracket with a built-in engine makes it possible to load different kinds of cargo on the roof -- the double floor has a compartment for skis (similar to a-class vehicles; sandwich construction) -- design: especially geared to the a-class mercedes ---

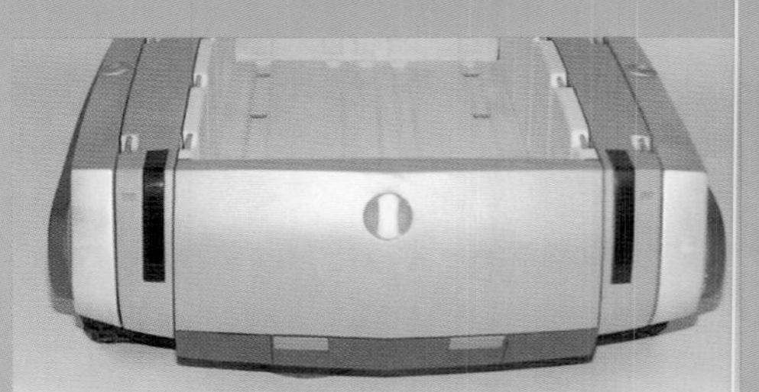

» croft «

» petrus gartler, roland lindner, volker pflüger «

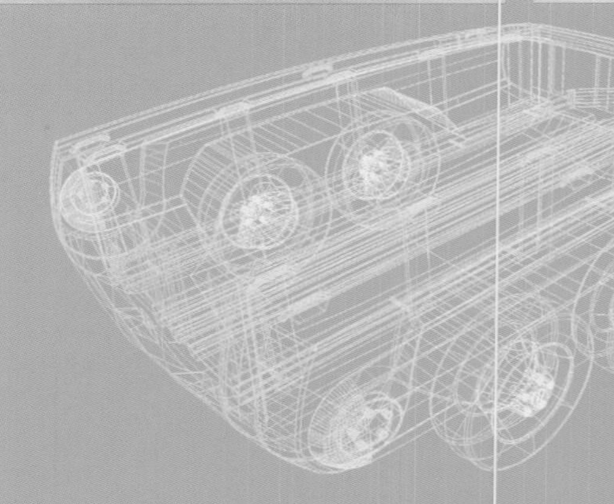

--- zielgruppe: - offroadbereich (baugewerbe, baustoffe, schüttgut etc..)
-- die innovation dieses fahrzeuges liegt in dem integrierten bügel (- als lastenheber schwenkbar, - als kipparm für die wanne, - als stützkonstruktion für die dachplane) ---

--- target group: - off-road use (construction companies, construction material, bulk cargo, etc.) -- the innovative element in this vehicle is the integrated bracket (- can be swiveled as a load lift, - as tilting mechanism for the loading compartment, - a supporting element for the canvas top) ---

» smart pack «

» christian becker, thomas binder, christoph gredler «

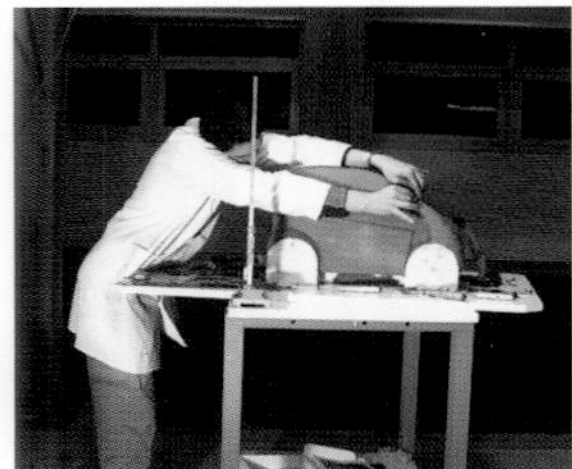

modellieren der 1:4 clayform ... *1:4 clay modeling ...*

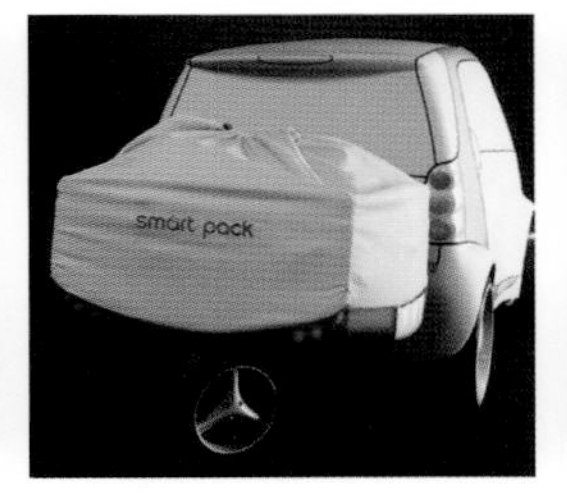

... ausgeklapptes backpack ... *folded-out backpack*

tape rendering

--- beim mcc-smart wird oft der platzmangel bei transporten kritisiert -- dafür wurde der "backpack" entwickelt -- beim laden grösserer stücke wird der backpack einfach heruntergeklappt und durch ein im nachlauf befindliches rad (mitlenkend) gestützt -- er wird an den 2 anschraubpunkten des mcc-smart, die für den ski- und fahrradträger vorgesehen sind, verankert ---

--- the mcc-smart has been criticized for lack of room to transport loads -- now a "backpack" was developed for it -- when transporting larger pieces, the backpack is simply folded down and is supported by a wheel (which helps with the steering) that is on the trailer -- it is anchored on the mcc-smart at 2 screw-on points, which are available for the ski and the bicycle racks ---

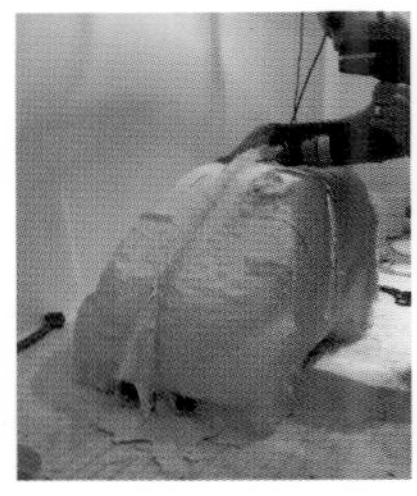

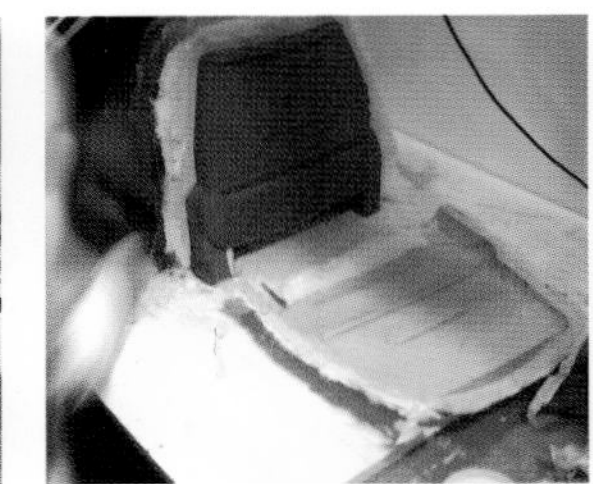

› › › › ›

...anfertigen + entformen des laminatnegatives ... *production and demolding of the laminated negative proof*

» audi st sport trailer «

» julian hönig, leonard natterer, christian zwinger «

--- an die audipalette angepasster reiseanhänger, konzipiert für den transport von sportgeräten -- das grundkonzept basiert auf einem mittels lamellendach verschliessbaren grundkörper und zwei höhenverstellbaren und umsteckbaren bügeln -- quer gesteckt dienen sie zur aufnahme von dachlasten, längs als planenunterkonstruktion -- die deichsel ist, um raum zu sparen, einschiebbar; der ganze hänger aufstellbar ---

--- a trailer suited for the audi model line and designed to transport sports equipment -- the concept is based on a basic body that can be closed by the lamellar roof and two brackets, whose height can be changed and which can be placed differently -- placed crosswise, they can be used to transport loads, lengthwise as a frame for the canvas top -- in order to save room, the shaft can be pushed inward; the entire trailer can be stood up on end ---

Foto: AUDI Design

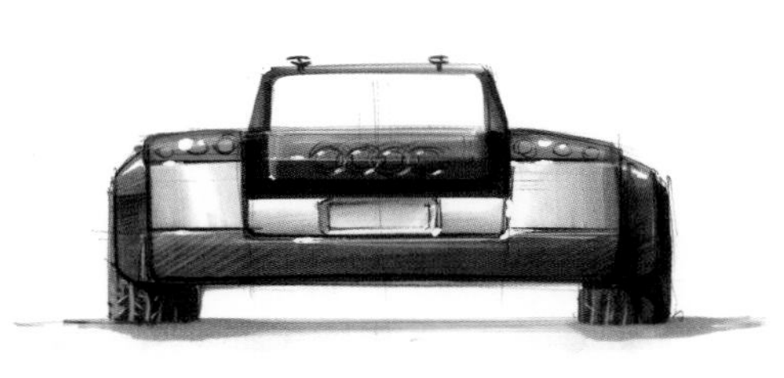

Foto: AUDI Design

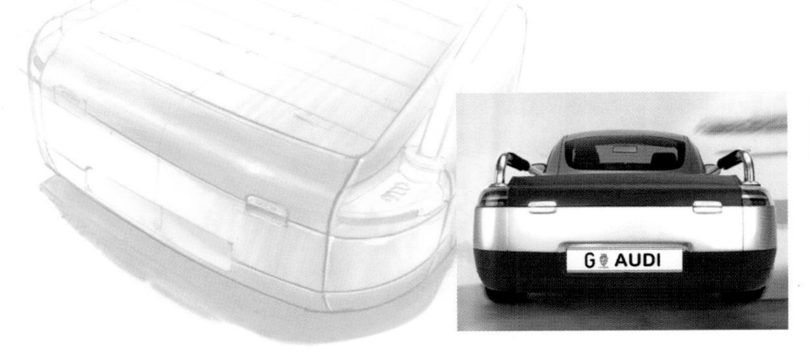

1999 (12)

projekt: » diplomarbeiten «
- mit div. kooperationspartnern
8. semester/sommer 1999

betreuer: gerhard h e u f l e r - gerald k i s k a - professoren für design -
georg w a g n e r - professor für engineering -
werner k l e i s s n e r - walter l a c h - modellbau –

themen: die themenbereiche des ersten diplomjahrganges spannen sich von low tech zu high tech, von der friedhofsurne bis zum digitalen notenständer, vom barrierefreien eissegler bis zur mobilen erste hilfe station. die projekte wurden zum teil in kooperation mit forschungsinstituten oder unternehmen durchgeführt, zum teil auch im ausland bearbeitet. alle diplomarbeiten wurden im rahmen der ersten grazer degree show 1999 den vertretern der designszene präsentiert.

project: » theses «
- with various collaborators
8th semester/spring 1999

advisors: gerhard h e u f l e r - gerald k i s k a - professors of design -
georg w a g n e r - professor of engineering -
werner k l e i s s n e r - walter l a c h - model building -

subject: the topics of the first theses run the gamut from low tech to high tech, from a burial urn to a digital music stand, from a barrier-free ice-sailboat to a mobile first-aid station. some of the projects were done in collaboration with research institutes or companies and some were worked on abroad. all the theses were presented to representatives of the design scene within the scope of the first graz degree show 1999.

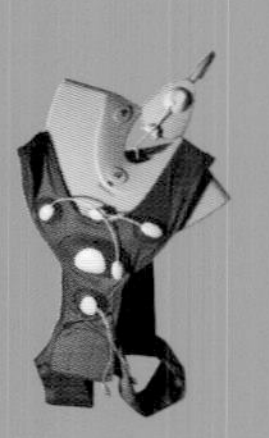

schlaflabor

iceslider

amoeba

pegasus

i.c. street

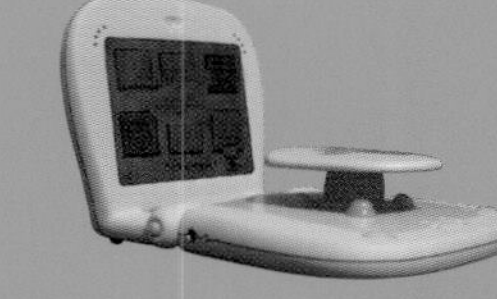
therapeutisches gerät

diplom-arbeit

de sign 

projektarbeiten des fh-studienganges industrial design-graz
projects of the fh-degree course industrial design-graz

1999
(12)

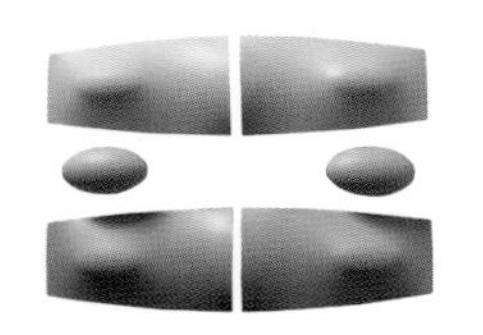

fasal-urne

nature design tv

oxymate

m.a.p

3ddd

digarranger

save unit

en 1999

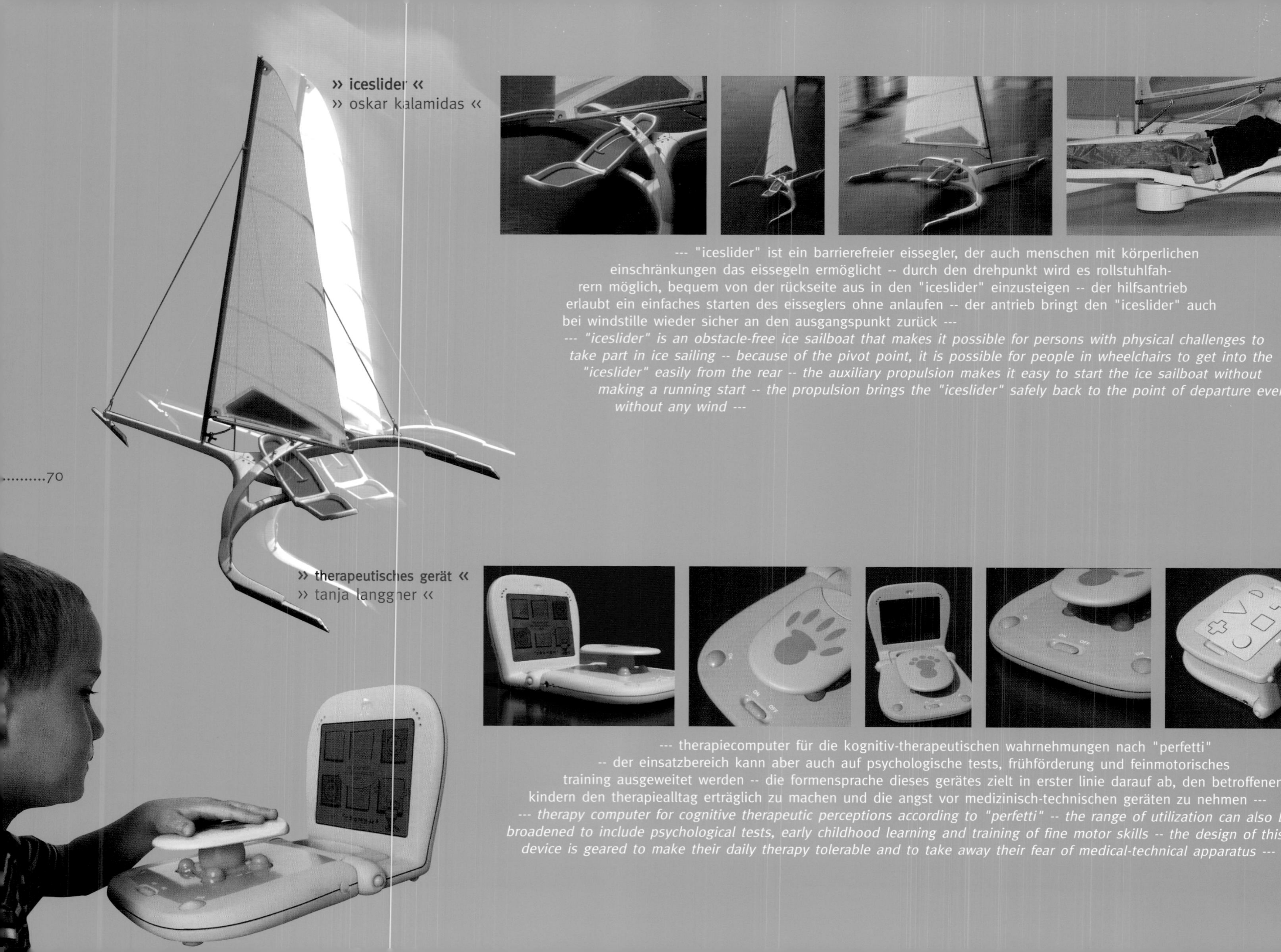

» iceslider «
» oskar kalamidas «

--- "iceslider" ist ein barrierefreier eissegler, der auch menschen mit körperlichen einschränkungen das eissegeln ermöglicht -- durch den drehpunkt wird es rollstuhlfahrern möglich, bequem von der rückseite aus in den "iceslider" einzusteigen -- der hilfsantrieb erlaubt ein einfaches starten des eisseglers ohne anlaufen -- der antrieb bringt den "iceslider" auch bei windstille wieder sicher an den ausgangspunkt zurück ---

--- "iceslider" is an obstacle-free ice sailboat that makes it possible for persons with physical challenges to take part in ice sailing -- because of the pivot point, it is possible for people in wheelchairs to get into the "iceslider" easily from the rear -- the auxiliary propulsion makes it easy to start the ice sailboat without making a running start -- the propulsion brings the "iceslider" safely back to the point of departure even without any wind ---

» therapeutisches gerät «
» tanja langgner «

--- therapiecomputer für die kognitiv-therapeutischen wahrnehmungen nach "perfetti" -- der einsatzbereich kann aber auch auf psychologische tests, frühförderung und feinmotorisches training ausgeweitet werden -- die formensprache dieses gerätes zielt in erster linie darauf ab, den betroffenen kindern den therapiealltag erträglich zu machen und die angst vor medizinisch-technischen geräten zu nehmen ---

--- therapy computer for cognitive therapeutic perceptions according to "perfetti" -- the range of utilization can also be broadened to include psychological tests, early childhood learning and training of fine motor skills -- the design of this device is geared to make their daily therapy tolerable and to take away their fear of medical-technical apparatus ---

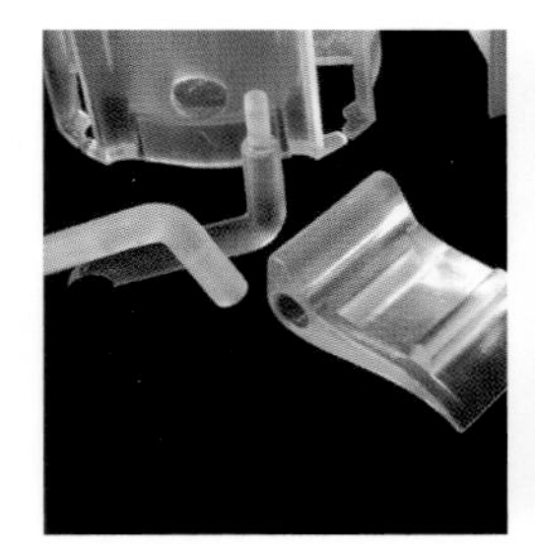

--- source

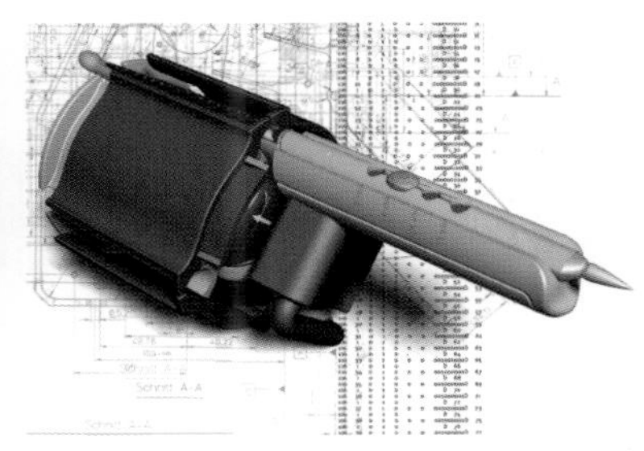

--- stylus

--- basisgerät *base station*

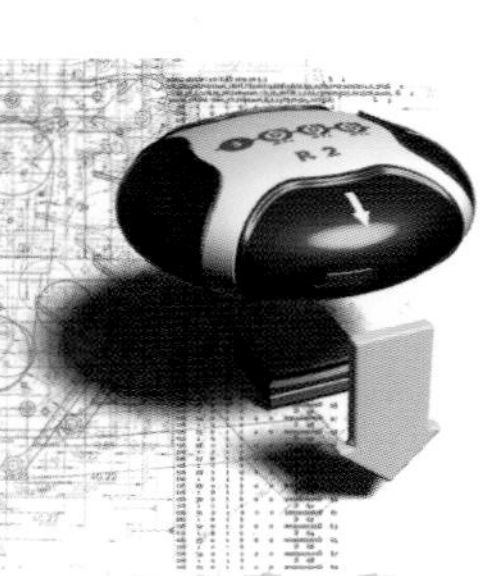

--- R2 receiver

» 3ddd «
» niki pelzl «

--- 3d desktop digitizer oder kurz "3ddd" genannt, ist ein mobiler 3d-scanner zum erfassen von räumlichen geometrien -- die 4 komponenten des gerätes sind: - das basisgerät, - die source, - der stylus (taster und somit empfänger), - r2 (der zweite empfänger in diesem system) -- dieses produkt nutzt elektromagnetische felder, um die position und die richtung eines vom ursprung entfernten objektes zu bestimmen ---

--- 3d desktop digitizer, called the "3ddd" for short, is a mobile 3d scanner that captures three-dimensional geometry -- the 3 components of the device are: - the basic device, - the source, - the stylus (sensor and thus receiver as well), - r2 (the second receiver in this system) -- this product uses electromagnetic fields in order to determine the position and the direction of an object, which is at a distance from the source ---

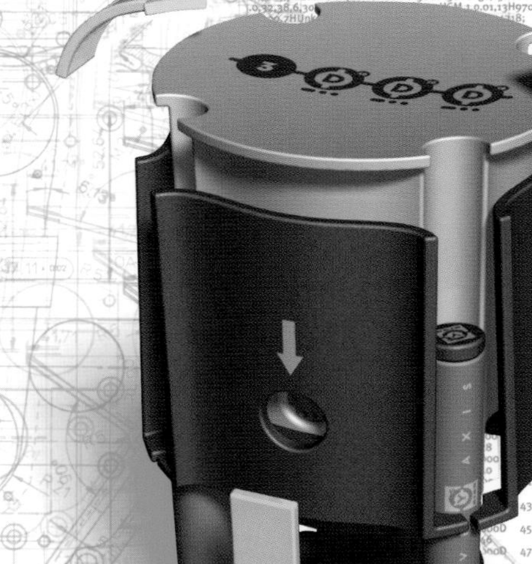

» oxymate «
» gernot puschner «

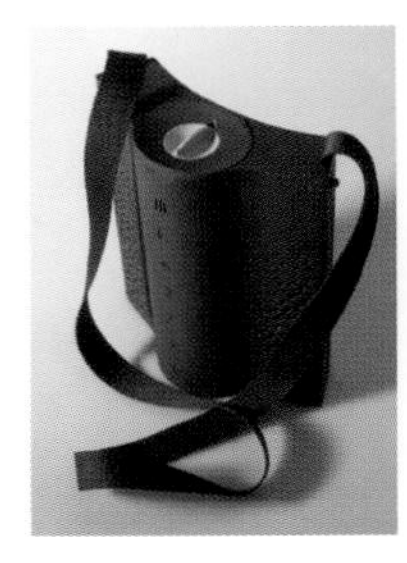

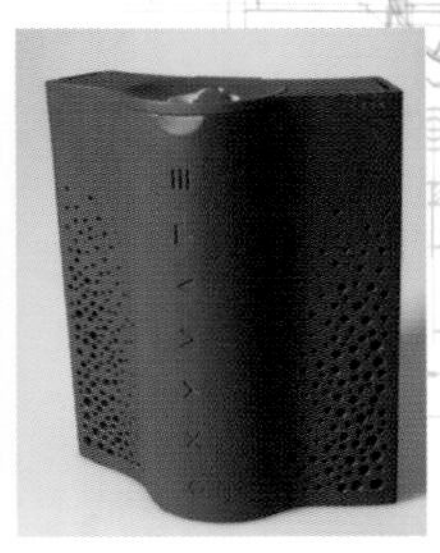

--- flüssigsauerstoffsysteme werden zur behandlung von lungenkrankheiten eingesetzt (leichtere atmung) -- besteht aus 2 komponenten: - 1. vorratsbehälter (station) als reservoir für die abfüllung von sauerstoff in den tragbaren behälter; der vorratsbehälter steht zu hause und wird regelmässig nachgefüllt - 2. tragbarer behälter (oxymate); der patient kann bis zu 8 stunden ausser haus gehen -- der behälter kann wahlweise mit einem schultergurt, einem gürtel oder einem rucksack getragen werden ---

--- liquid oxygen systems are used to treat lung diseases (for easier breathing) -- it consists of two components: - 1. storage tank (station) used as a reservoir to fill the oxygen into the portable container; the storage tank is stored at home and is regularly refilled - 2. portable container (oxymate); the patient can leave the house for up to 8 hours -- the container can be carried either with a shoulder strap, a belt, or a backpack ---

» amoeba «
» klaudia takats «

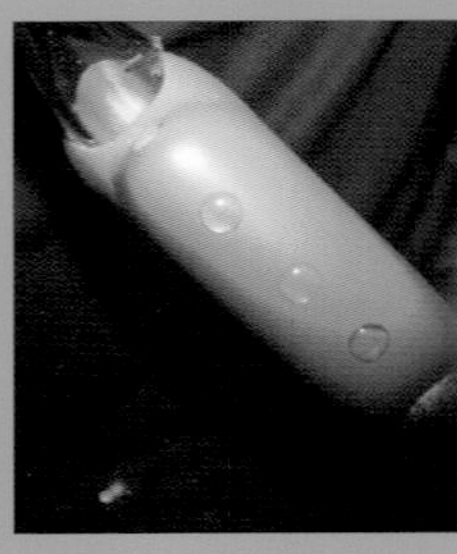

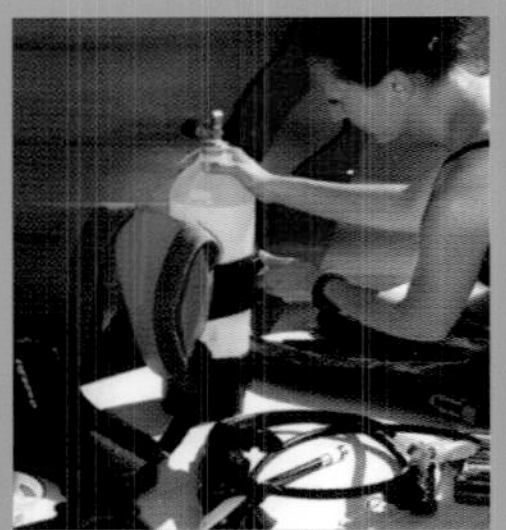

--- "amoeba" ist eine art wing-jacket, das aus zwei voneinander unabhängigen teilen besteht: - dem rucksack, der zur befestigung der sauerstoffflasche dient und - die tarierweste ---
--- "amoeba" is a kind of wing jacket that consists of two independent parts: - the backpack, which is used to fasten the oxygen bottle and - the buoyancy compensator vest ---

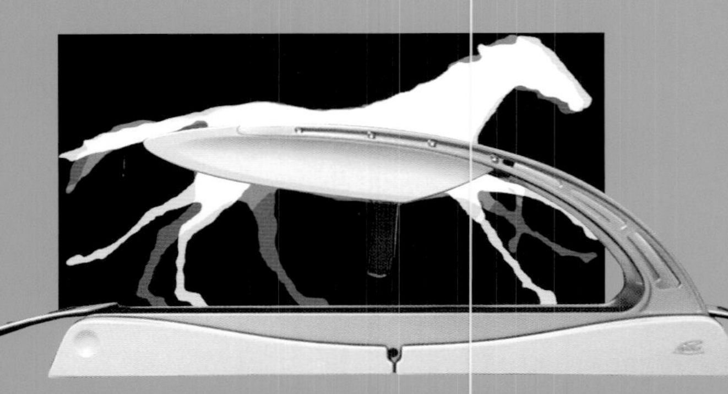

--- pferde-trainer, - förderband aus gewebegummi, - seitliche hohlrahmenträger -- die rampen sind drehbar gelagert, um auch bei der maximalen neigung von 14% noch begehbar zu sein -- der träger begrenzt mit dem flügel den arbeitsbereich des pferdes und hat auch das sicherheitssystem integriert (ein gurtgeschirr) ---
--- horse trainer, - conveyor belt made of rubber fabric, - side hollow frame support -- the ramps are set up so that they can be pivoted in order to be accessible even when set at the maximum incline of 14% -- the support confines the horse's working area with a wing, and it also has the safety system (harness) integrated into it ---

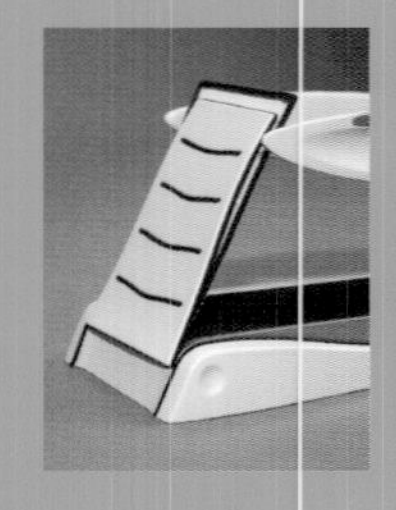

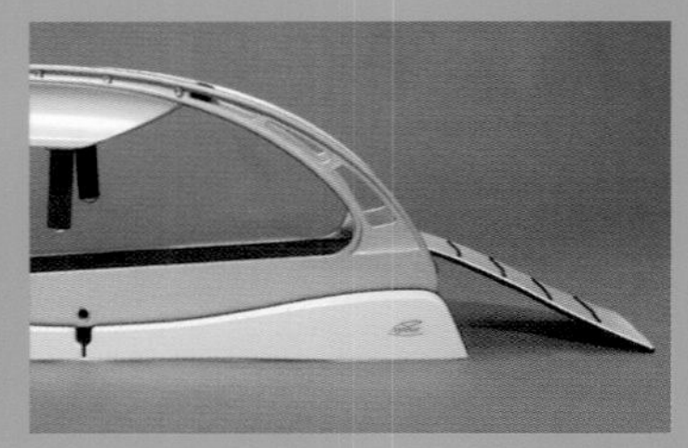

» pegasus «
» david stelzer «

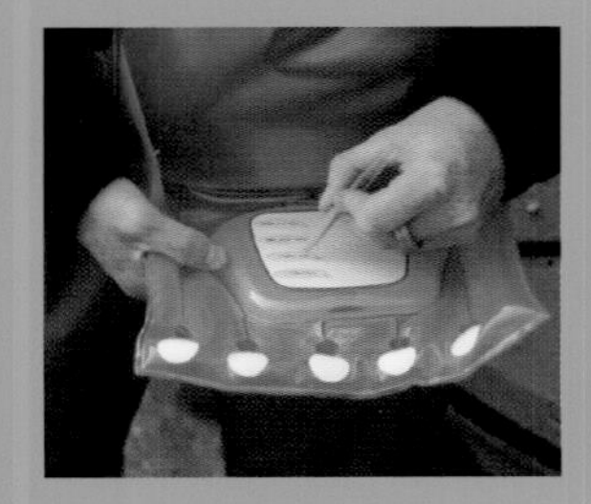
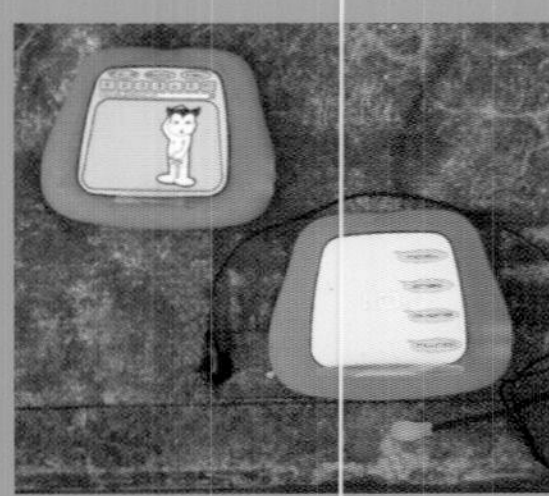

» i.c. street «
» sylvia feichtinger «

--- inflate communicationbag for street culture ist ein bag mit kommunikatorteil: das produkt besitzt vier icons: - mail, - inet, - phone, - music ---
--- inflatable communication bag for street culture is a bag with a communication component: the product has four icons: - mail, - internet, - phone, - music ---

--- das bioabbaubare plastiksurrogat fasal verbindet die positiven eigenschaften des werkstoffs holz mit der leichten verarbeitbarkeit von kunststoff -- die urne setzt sich aus flächigen elementen zusammen, die ein ellipsoid umfassen, in dem sich die asche befindet -- an der oberseite formen die flächen einen trichter aus, die dem wurzelstock eines baumes raum bieten ---

--- the biodegradable plastic substitute fasal combines the positive qualities of wood with the easy workability of plastic -- the urn consists of flat elements that surround an ellipsoid, which contains the ashes -- at the top, the flat surfaces form a funnel, which has enough space for the root ball of a tree ---

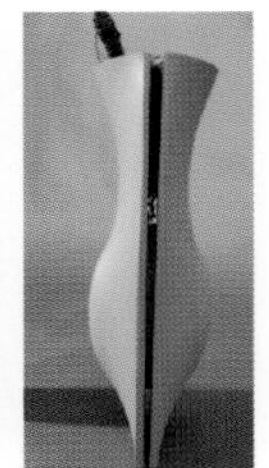

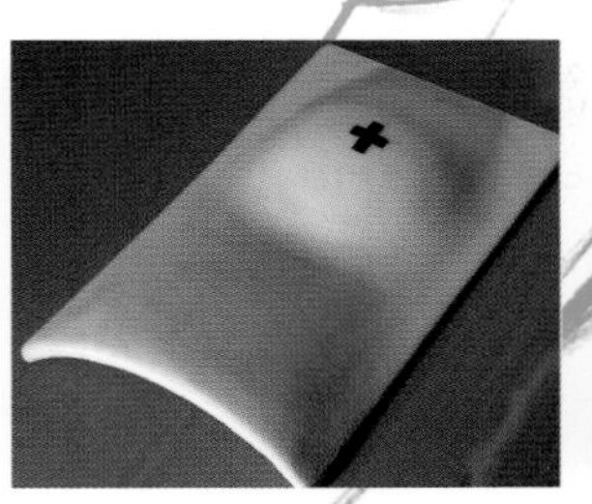

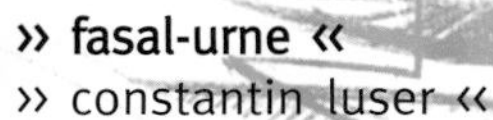

» fasal-urne «
» constantin luser «

» nature design tv «
» johannes scherr «

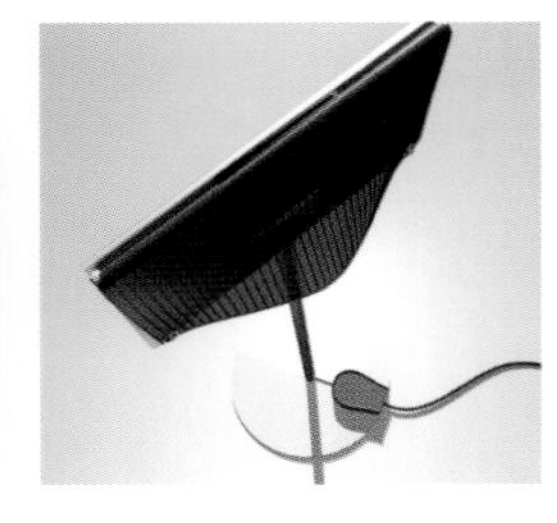

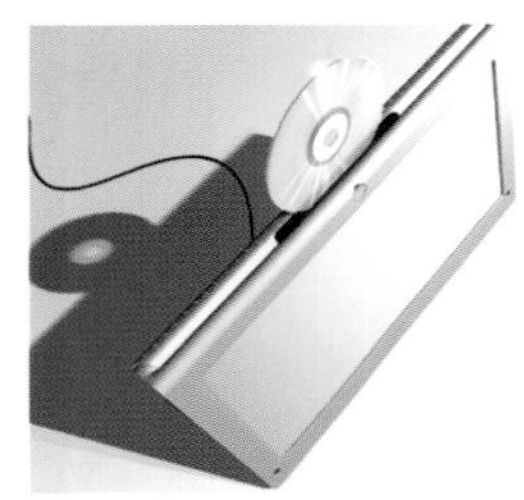

--- "nature design tv" ist das ergebnis eines ökologieorientierten designs mit den schwerpunkten langlebigkeit, geringer energieverbrauch, persönlichkeit und tragbarkeit -- ein erscheinungsbild, das aus den eigenschaften der verwendeten materialien resultiert: eine flexible abdeckung wird zum rücken gefaltet und entwickelt sich zur stehhilfe ---

--- "nature design tv" is the result of an ecological design focused on long durability, low energy use, personality, and portability -- its appearance is a result of the qualities of the utilized materials: a flexible covering is folded to form a back and becomes a support to keep it upright ---

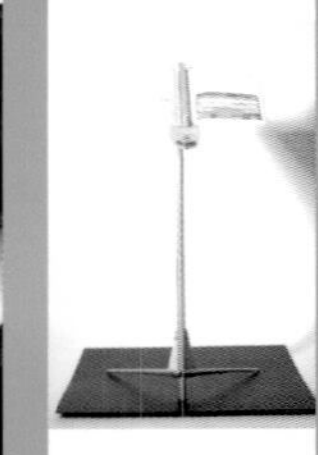

» digarranger «

» harald auer «

--- der "digitale notenständer" ist die vereinigung von einem klassischen notenständer mit einem personalcomputer -- zwei flexible, ausrollbare polymerfoliendisplays reduzieren den papiernotenaufwand auf digitale seiten -- eingebautes, digitales stimmgerät, metronom und schnittstellen zu externen geräten -- bearbeitung der noten mittels digitalem stift ---

--- the "digital note stand" is a combination of a classic note stand and a personal computer -- two flexible polymer foil displays, which can be rolled out, are digital pages used instead of sheet music -- built-in, digital tuning device, metronome, and interfaces to external devices -- annotations and revision of the notes are made with a digital pen ---

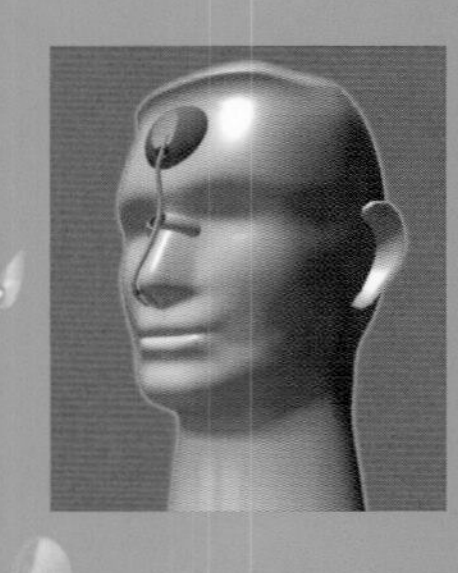

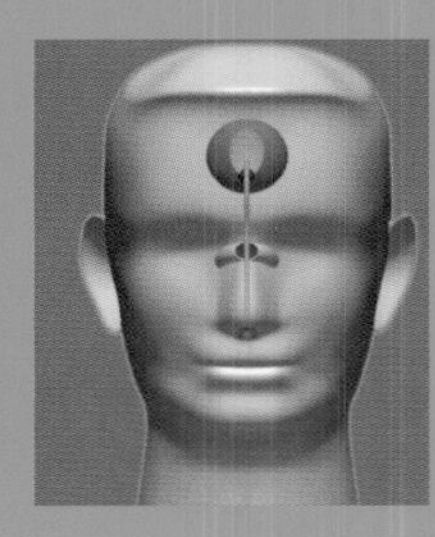

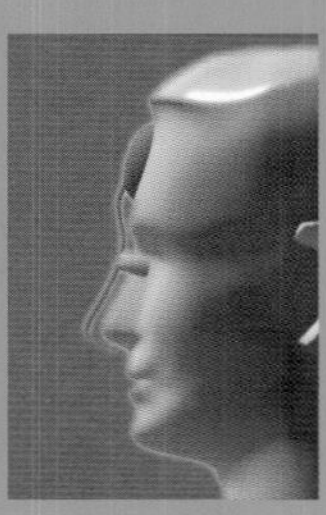

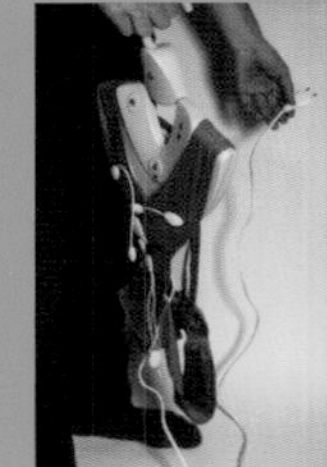

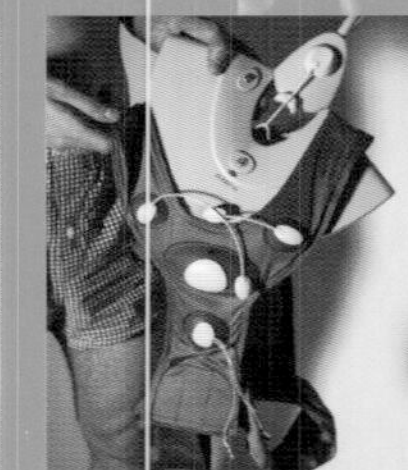

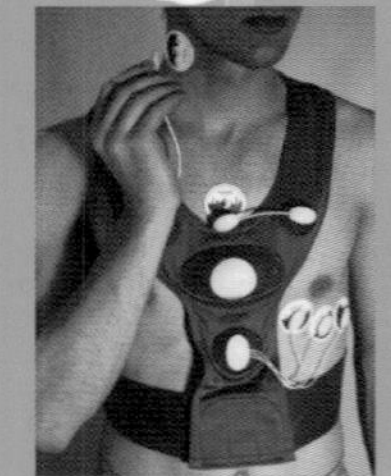

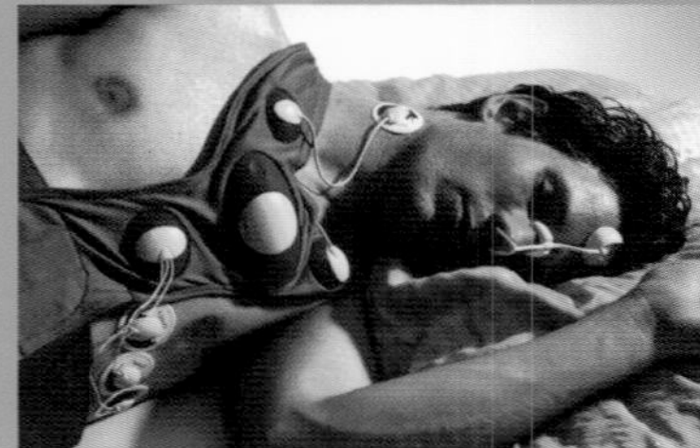

» schlaflabor «

» silke spath «

--- "sleeplab" ist ein kleines labor zur schlafaufzeichnung und wurde in zusammenarbeit mit dem joanneum research für den ambulanten einsatz konzipiert -- der vorteil gegenüber einem stationären schlaflabor liegt darin, dass der patient zu hause die daten von ekg, puls, atmung und körperbewegung aufzeichnen kann -- der patient kann sich unmittelbar vor dem schlafengehen die "jacke" mit den elektroden anziehen -- am nächsten tag legt der patient "sleeplab" wieder ab und überträgt per knopfdruck die gesammelten daten zu seinem arzt ---

--- "sleeplab", a small lab to record sleep patterns, was designed in collaboration with joanneum research for ambulatory use -- the advantage as opposed to a stationary sleep lab is that the patient can record the ekg, pulse, breathing, and body movement data at home -- the patient puts on the "jacket" with the electrodes directly before going to bed -- the next day, the patient takes off the "sleeplab" and, at the push of a button, transmits the recorded data to his/her doctor ---

›› m.a.p. ‹‹
›› christian winkler ‹‹

--- "mobile access point" ist ein interaktiver portabler reiseführer für den städtischen bereich -- digitaler informationsträger, interaktive landkarte und eingebauter gps-empfänger ---

--- "mobile access point" is an interactive, portable urban travel guide -- digital data medium, interactive map, and built-in gps receiver ---

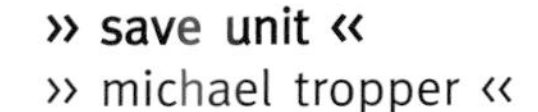

›› save unit ‹‹
›› michael tropper ‹‹

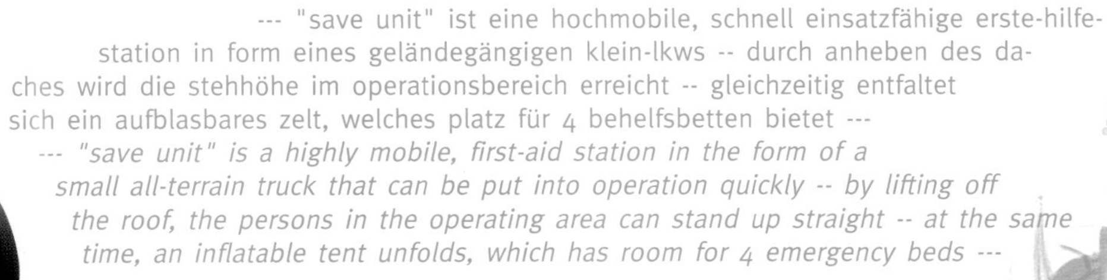

--- "save unit" ist eine hochmobile, schnell einsatzfähige erste-hilfe-station in form eines geländegängigen klein-lkws -- durch anheben des daches wird die stehhöhe im operationsbereich erreicht -- gleichzeitig entfaltet sich ein aufblasbares zelt, welches platz für 4 behelfsbetten bietet ---

--- "save unit" is a highly mobile, first-aid station in the form of a small all-terrain truck that can be put into operation quickly -- by lifting off the roof, the persons in the operating area can stand up straight -- at the same time, an inflatable tent unfolds, which has room for 4 emergency beds ---

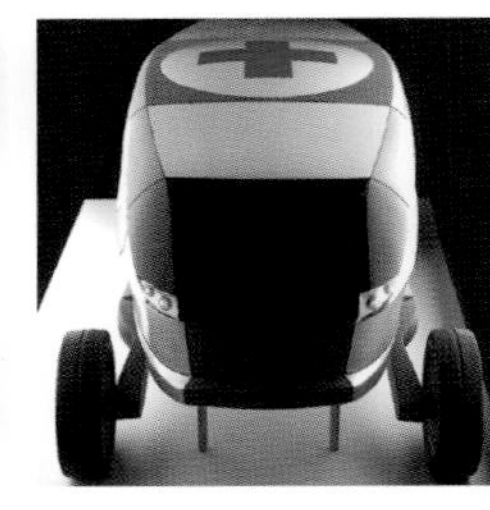

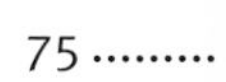

1999/2000 (13)

projekt: » rednerpulte wifi 2005 «
- in kooperation mit dem wifi steiermark
3. semester/winter 1999/2000

betreuer: gerhard h e u f l e r - gerald k i s k a - professoren für design -
georg w a g n e r - professor für engineering -
werner k l e i s s n e r - walter l a c h - modellbau –

thema: das wirtschaftsförderungsinstitut spricht verstärkt als zielgruppe das gehobene management an. das soll sich auch in der produktsprache einer neuen generation von rednerpulten ausdrücken: high tech ausstrahlung und innovatives erscheinungsbild sind gefragt! einsatzbereiche: kongresse, bankette und seminare. zu den weiteren anforderungen gehören: höhenverstellbarkeit, leichte transportfähigkeit und optimale ergonomische bedingungen.

project: » lecterns wifi 2005 «
- in collaboration with wifi steiermark
3rd semester/fall 1999/2000

advisors: gerhard h e u f l e r - gerald k i s k a - professors of design -
georg w a g n e r - professor of engineering -
werner k l e i s s n e r - walter l a c h - model building -

subject: the activities of the institute for economic development (wifi) are primarily directed toward upper management as a target group. this fact should find expression in a new generation of lecterns – a high tech feel and an innovative appearance are desirable! possibilities for use are: conferences, banquets, and seminars. these lecterns should be height-adjustable, easily transportable, and offer optimal ergonomics.

rednerpulte

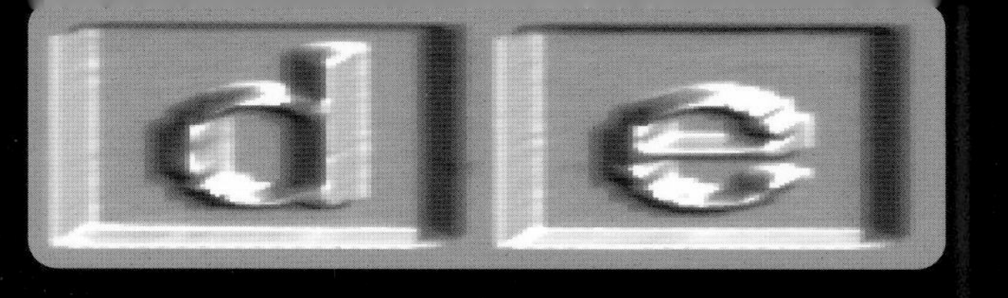

sign

projektarbeiten des fh-studienganges industrial design-graz

projects of the fh-degree course industrial design-graz

1999/2000
(13)

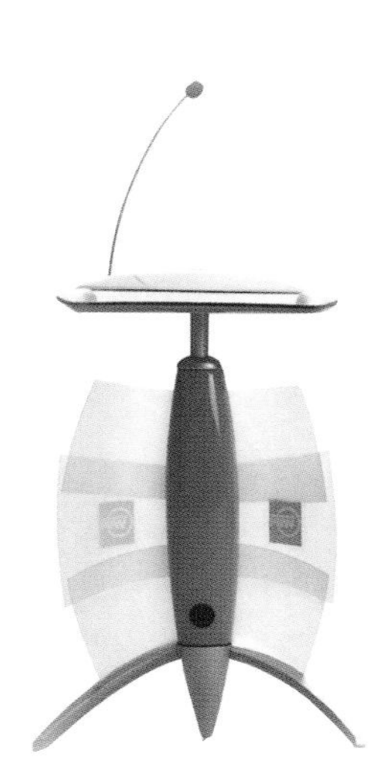

wifi 2005

» robert hitthaler «

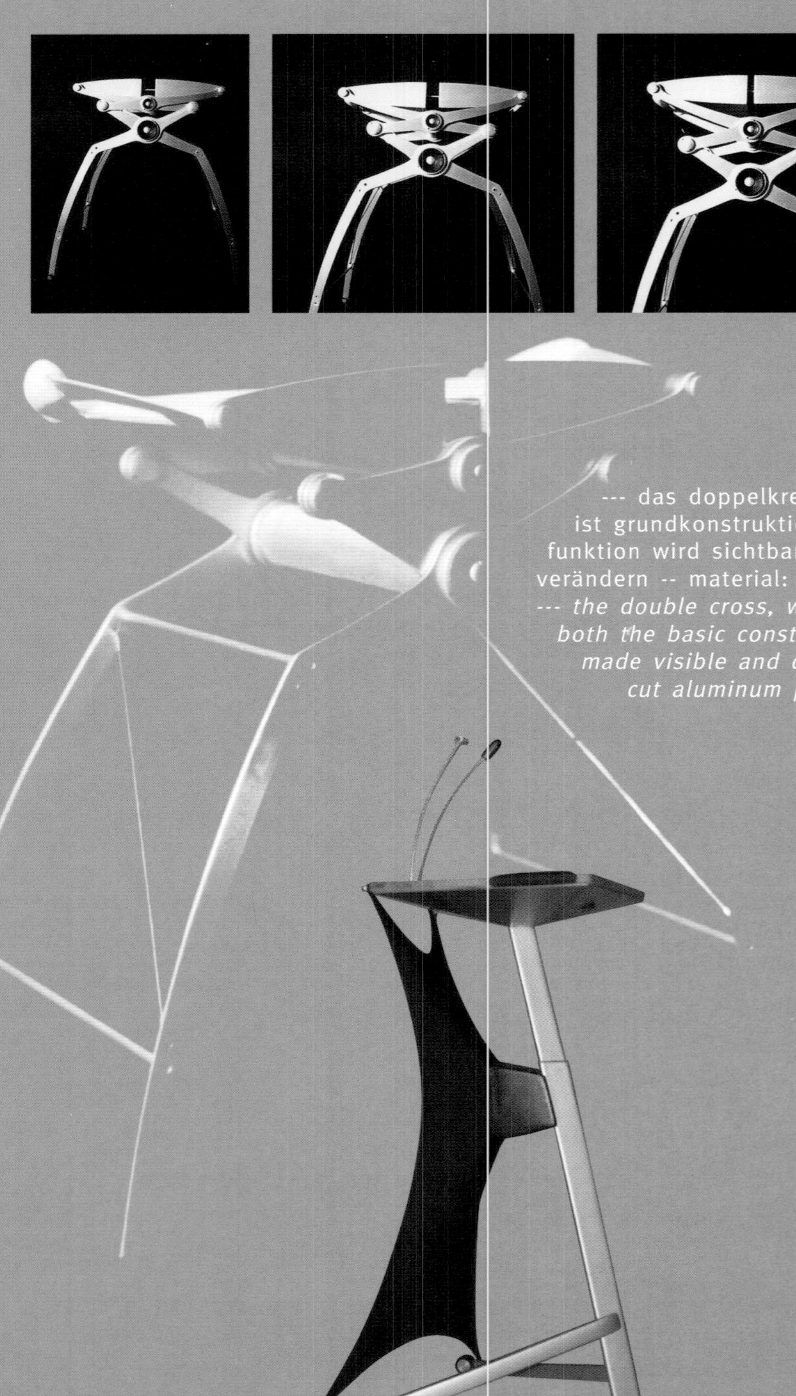

--- das doppelkreuz, das das prinzip der höhenverstellung darstellt, ist grundkonstruktion und hauptgestaltungsmerkmal zugleich -- die funktion wird sichtbar gemacht und kann den charakter des pultes verändern -- material: - lasergeschnittene alubleche ---

--- the double cross, which illustrates the principle of height adjustment, is both the basic construction and the main design characteristic -- the function is made visible and can change the character of the lectern -- material: - laser-cut aluminum plate ---

--- die asymmetrie verleiht dem pult eine offene form und lässt es mit dem publikum kommunizieren -- material: - elastischer mikrofaserstoff, - aluminiumlegierung ---

--- asymmetry gives the lectern an open design and lets it communicate with the public -- material: - elastic microfiber material, - aluminum alloy ---

» gabriele bruner «

» andrea platzer «

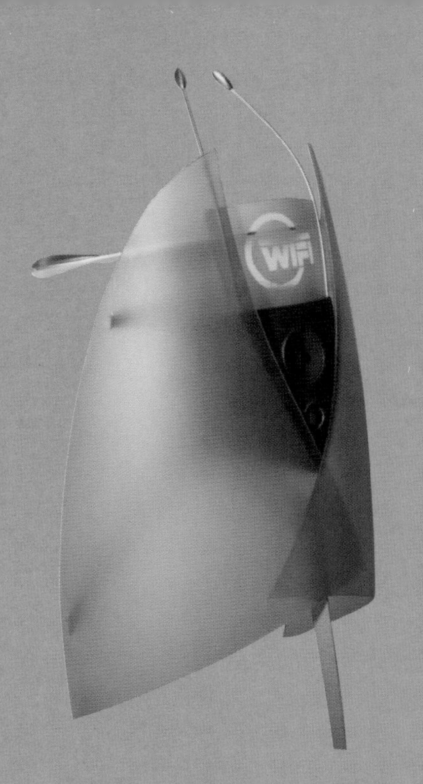

--- prinzip: - integrierte technik, - ergonomischer aufbau, - psychologische schutzfunktion, - transportabel -- grundidee: das ineinander verschieben zweier transluzenter schalen um eine achse bewirkt ein wandelbares erscheinungsbild -- materialwahl: alu (matt) ---

--- principle: - integrated technology, - ergonomic structure, - psychological protective function, - transportable -- basic idea: shifting two translucent shells into one another around an axis results in a changeable appearance -- material: aluminum (mat) ---

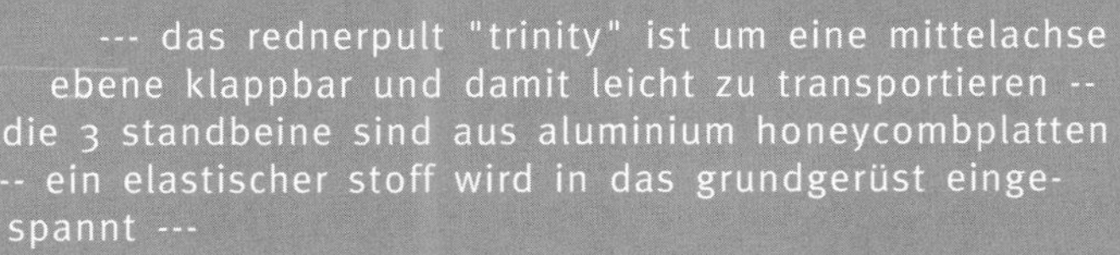

--- das rednerpult "trinity" ist um eine mittelachse zu einer ebene klappbar und damit leicht zu transportieren -- die 3 standbeine sind aus aluminium honeycombplatten -- ein elastischer stoff wird in das grundgerüst eingespannt ---

--- the "trinity" lectern has a middle axis and can be folded down to one surface so that it is easy to transport -- the three legs are made of aluminum honeycomb plates -- an elastic material is placed into the basic framework ---

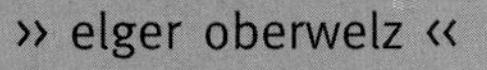

» elger oberwelz «

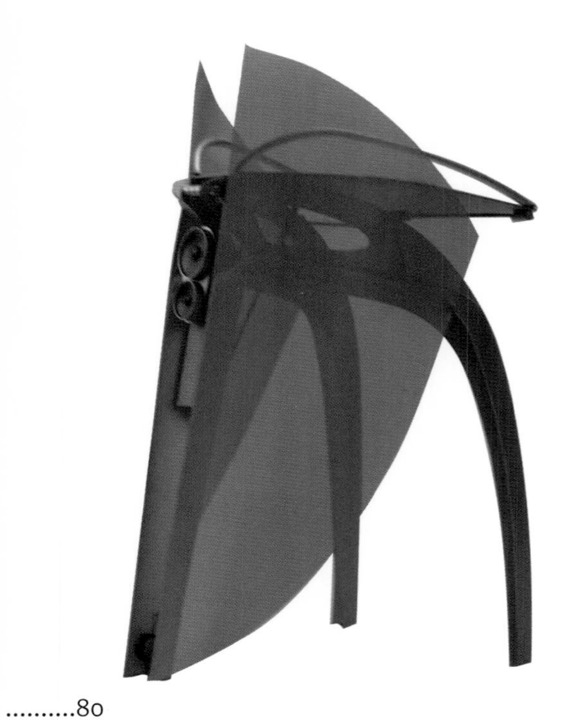

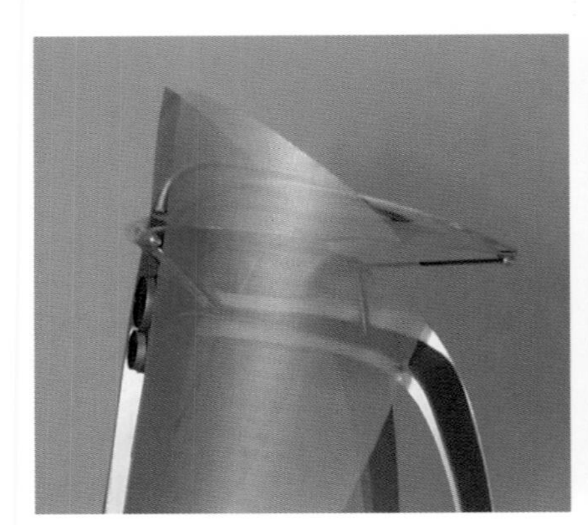

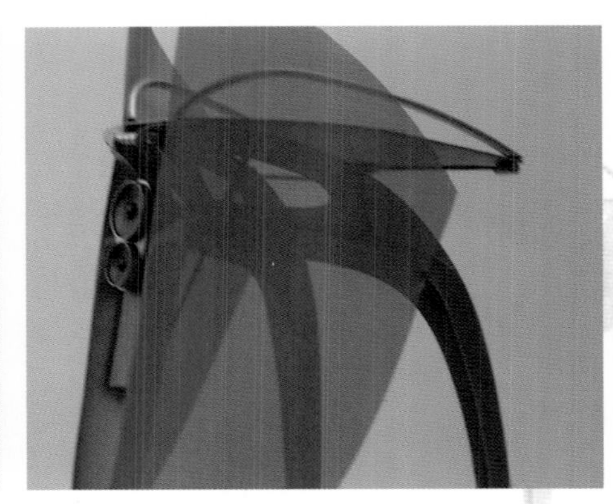

--- durch den konsequenten einsatz von metall wirkt das pult abstrakt und imaginär -- das pult ist klappbar und lässt sich durch ein ausfahrbares rad einfach transportieren -- schilder aus lochblech: wirkung transparenz ---

--- through consistent use of metal, the podium has an abstract and imaginary effect -- it can be folded and easily transported by means of a retractable wheel -- plates made of perforated metal sheets: transparent effect ---

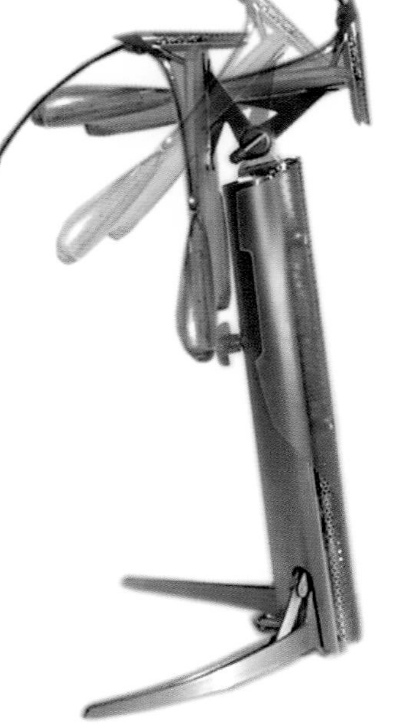

--- grundgedanke: - grosse mobilität und flexibilität, - sichtschutz, - an die bedürfnisse des redners anpassbar -- das rednerpult kann zu einer kompakten einheit zusammengeklappt werden -- standbeine klappen hoch, kleine transportrollen kommen zum vorschein -- die pultfläche ist stufenlos neigbar ---

--- basic idea: - high mobility and flexibility, - privacy screen, - can be adapted to the speaker's needs -- the lectern can be folded up to form a compact unit -- legs can be folded up, small rollers for transport appear -- infinitely variable tilting of the surface of the podium possible ---

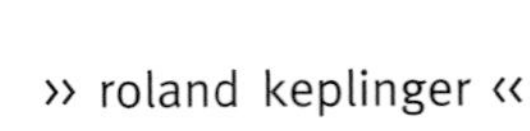

›› roland keplinger ‹‹

» jan mathis «

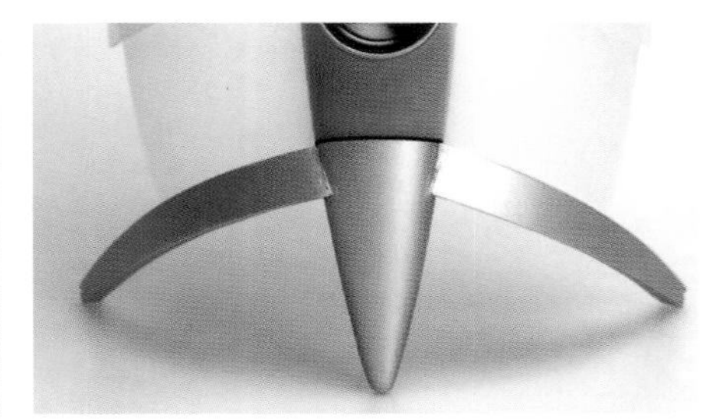

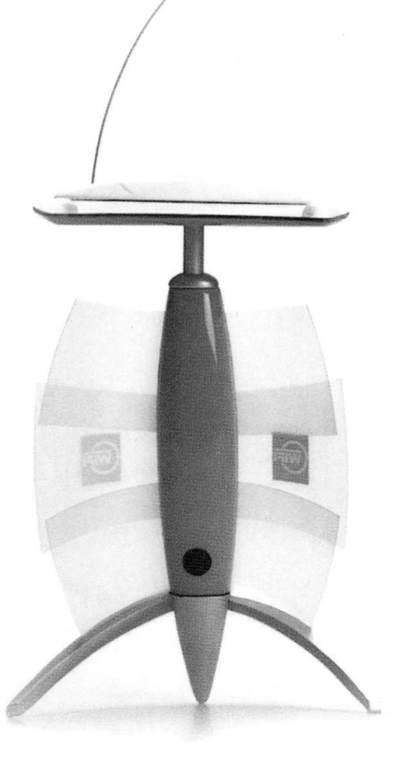

--- zentrale säule, höhenverstellbar mittels gasdruckfeder, sichtschutz bestehend aus drei schuppen, die über elastische bänder gekoppelt sind und sich bei verstellung der höhe aus- bzw. ineinander verschieben -- material: - acrylglas, - aluminium ---

--- central column, hydraulic cylinder makes variable heights possible, privacy screen made up of three rounded units, which are connected by elastic cords and which pull apart or are pushed together when the height is changed -- materials: - acrylic glass, - aluminum ---

--- die pultfläche kann auf beiden seiten radial ausgezogen werden -- anschlüsse und lautsprecher befinden sich im keil zwischen den beiden pultflächen ---

--- the lectern surface can be extended radially on both sides -- electrical connections and loudspeakers are in the wedge between both lectern surfaces ---

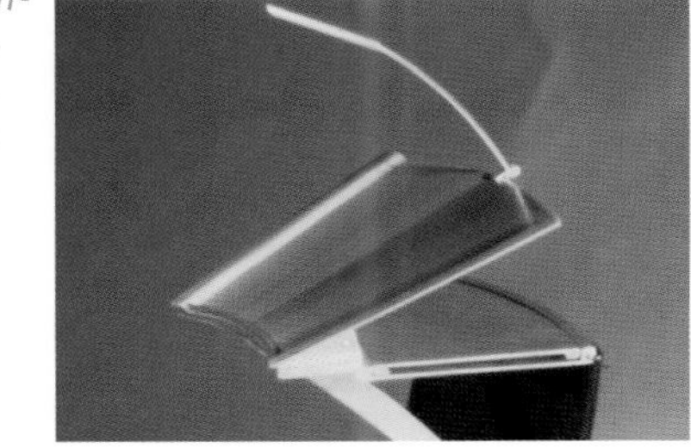

» rainer trummer «

1999/2000 (14)

projekt: » nahverkehr-schienenfahrzeuge «
7. semester/winter 1999/2000

betreuer: dirk s c h m a u s e r (porsche design) - gastprofessor
georg w a g n e r - professor für engineering
werner k l e i s s n e r + walter l a c h - modellbau

thema: die umwelt- und gesundheitsbelastung durch den individualverkehr wird immer kritischer. ziel muss es daher sein, den immer intensiver werdenden nah- und regionalverkehr von der strasse verstärkt auf die schiene zu verlegen. zielgruppen sind Menschen, die das verkehrsmittel beruflich oder privat täglich oder mehrmals wöchentlich nutzen. hauptkriterien: schnelles und bequemes ein- und aussteigen, hoher komfort und sicherheit.

project: » local transport – rail vehicles «
7th semester/fall 1999/2000

advisors: dirk s c h m a u s e r (porsche design) - guest professor
georg w a g n e r - professor of engineering
werner k l e i s s n e r + walter l a c h - model building

subject: environmental pollution and health hazards caused by private transport are becoming more and more critical. therefore, it must be one of our goals to shift local and regional transport from the road to rail. target groups are people, who could use public transportation daily or several times per week for commuting to work and for personal errands. the main criteria are quick and easy boarding and getting off, comfort, and safety.

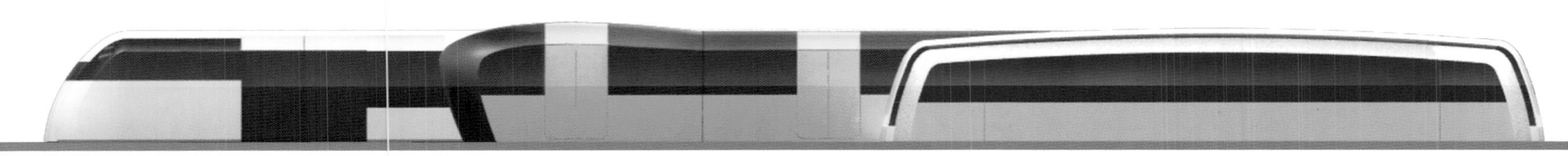

» mirai « » airportshuttle « » business-shuttle «

nahverkehr-schi

sign

projektarbeiten des fh-studienganges industrial design-graz

projects of the fh-degree course industrial design-graz

1999/2000
(14)

enenfahrzeuge

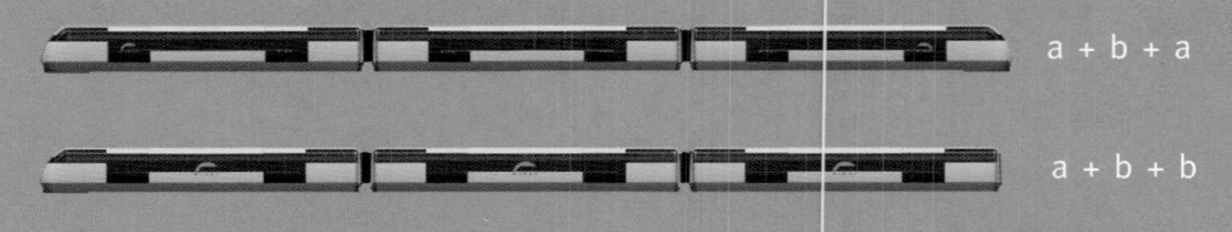

--- typologie: a = endwaggon -- b = standardwaggon ---

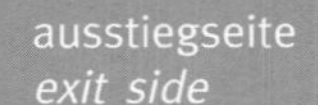

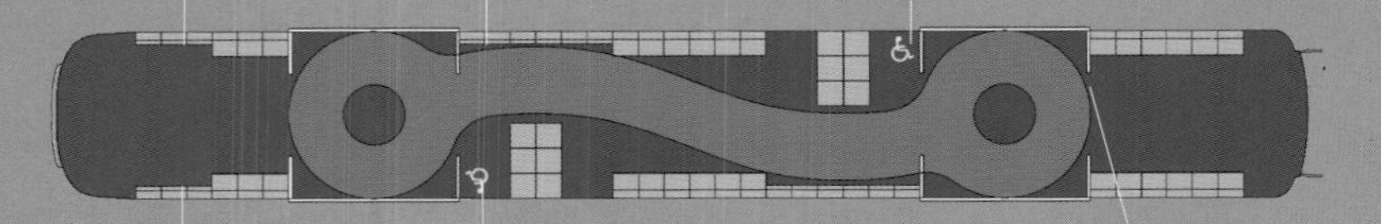

MIRAI

--- der zug ist für 2 grundsätzliche systeme von bahnlinien konzipiert: - a) ringlinie, - b) geradlinie -- je nach kundenwunsch wird die garnitur daher symmetrisch oder asymmetrisch montiert -- die eingangsbereiche werden durch kreiselemente im boden betont -- im mittelteil des waggons wird der "tunnel-effekt" bestehender züge durch schwünge in der bodengraphik und quergestellte sitzelemente vermieden ---

--- the train is designed for 2 basic kinds of railway lines: - a) ring lines, - b) straight lines -- depending on the customer's wish, the railway cars are installed symmetrically or asymmetrically -- the maintenance areas are emphasized by means of circular elements in the floor -- in the central part of the car, the "tunnel effect" of currently existing trains is avoided by sweeps in the floor graphics and seating components placed at right angles ---

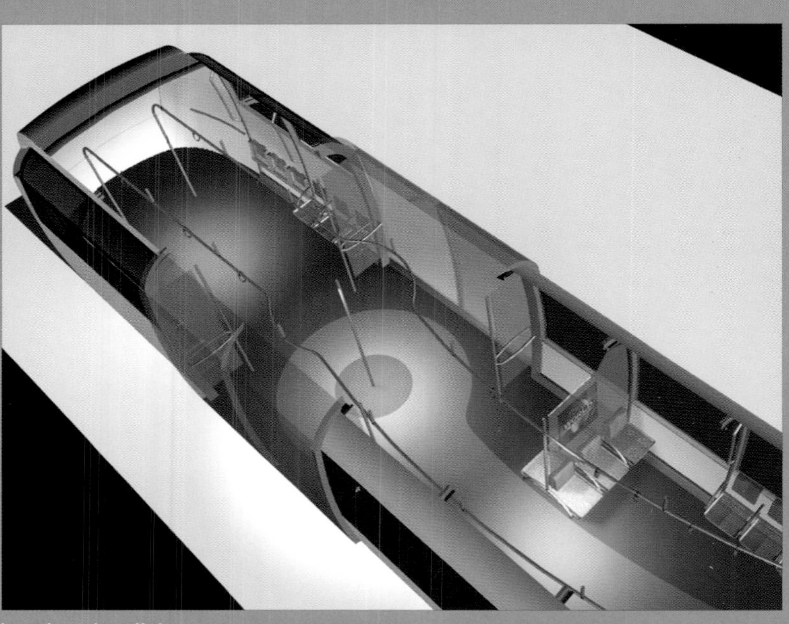

interior-visualisierung

...klappbare sitze, auch als steh-sitz-hilfe verwendbar ---

...foldable seats, which can also be used as an aide when sitting down or standing up ---

» mirai «

» christian becker, thomas binder, christoph gredler «

sketch

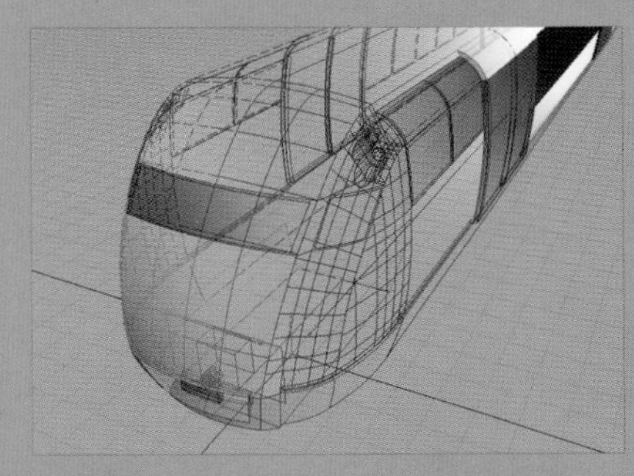

alias wireframe

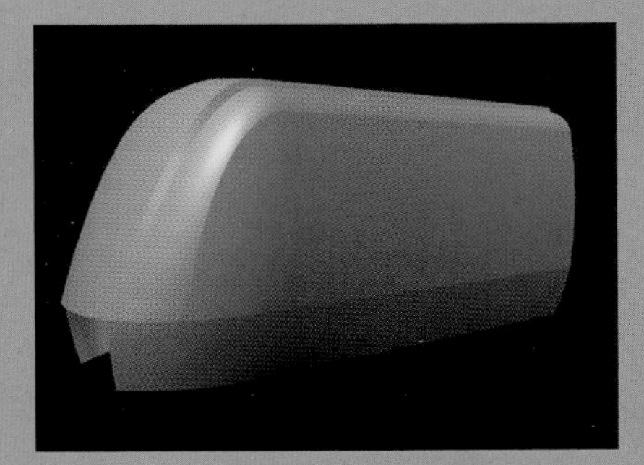

alias surface modeling

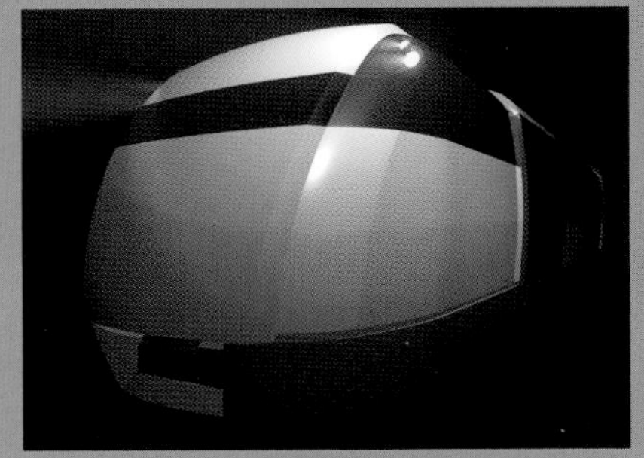

alias rendering

disualization/visualisierung team mirai

klappbarer rückenteil *foldable backrest*

--- interieur: business-office atmosphäre: klappbare sitzrücklehnen ermöglichen einen wechsel zwischen 2er- und 4er-gruppen mit tisch ---

--- interior: business office atmosphere: foldable seat backs make it possible to vary between 2-person groups and 4-person groups with table ---

rückenteil in fahrtrichtung
seat back in traveling direction

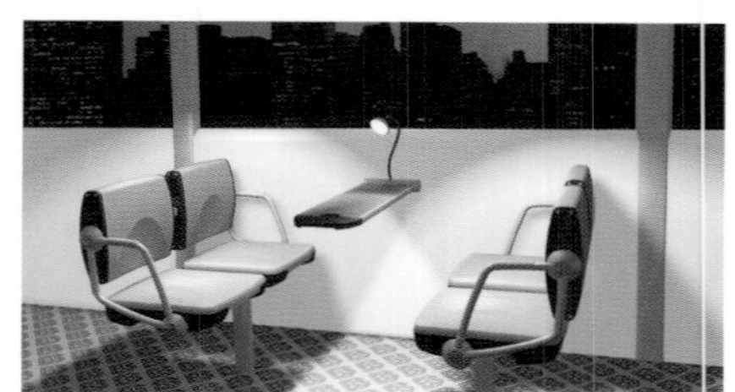

rückenteil gegen die fahrtrichtung
seat back opposed to traveling direction

» business-shuttle «

» julian hönig, roland lindner, volker pflüger «

--- konzept: - privatwirtschaftlich betriebenes shuttle, im öffentlichen u-bahnnetz, - als serviceleistung von grösseren konzernen für ihre mitarbeiter, - fahrerloser betrieb -- exterieur: aluminium spaceframekonstruktion als träger für glas und seitenpaneele sowie klimaaufsatz ---

--- concept: - privately operated shuttle within the public subway network, - as a service by large companies for their employees, - driverless operation -- exterior: aluminum spaceframe construction supporting glass and side panels, as well as the top-mounted air-conditioning unit ---

grafikkonzept:

graphics concept:

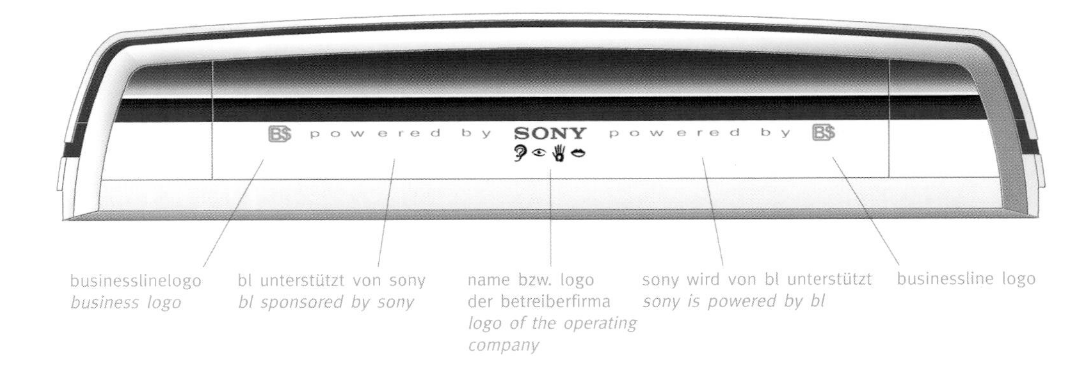

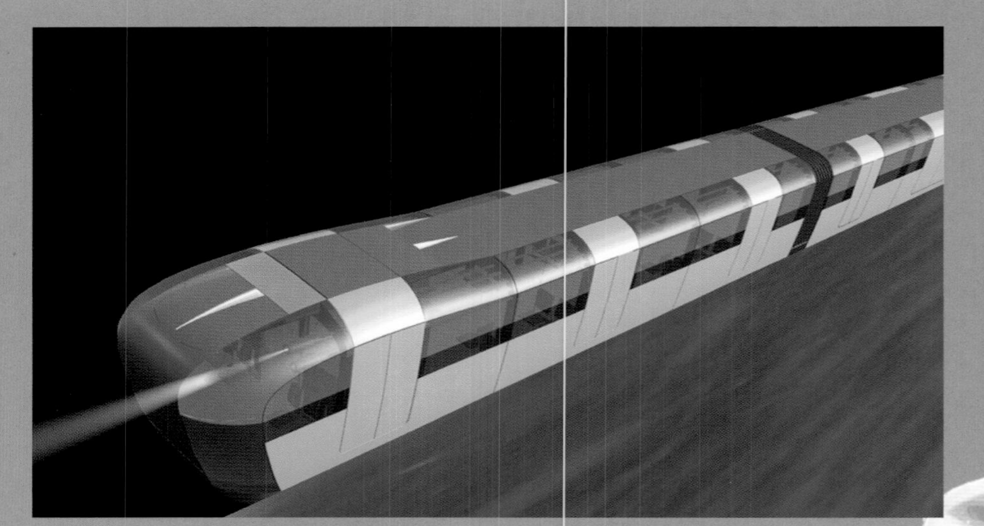

alias visualisierung

--- die bestuhlung ist im gesamten interieur gleich -- jeweils 2er- oder 3er-gruppen ohne verbindung zum boden, an den querstangen befestigt -- bei den 3er-gruppen ist jeweils ein stuhl drehbar gelagert und zu der bevorzugten sitzordnung zuwendbar ---

--- the seats are the same throughout the entire interior -- 2-person or 3-person groups are attached to horizontal bars and are not connected to the floor -- in the 3-person groups, one seat can be pivoted so that it can provide the desired seat configuration ---

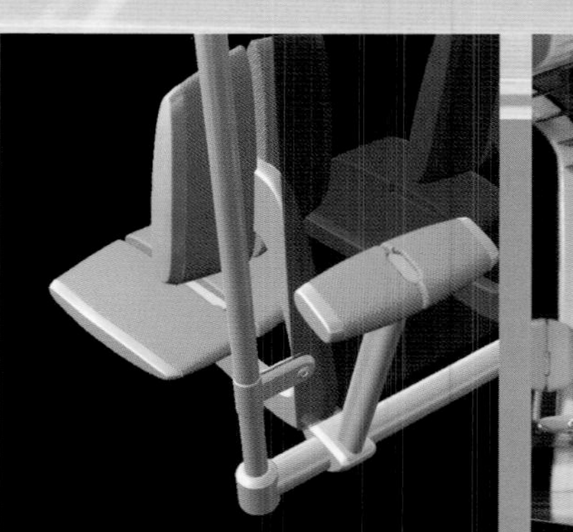

modell 1:5
model 1:5

» sputnik «

» leonard natterer, martin wöls, christian zwinger «

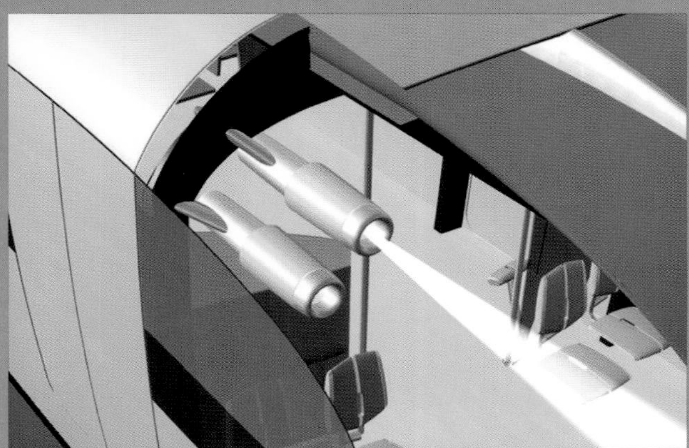

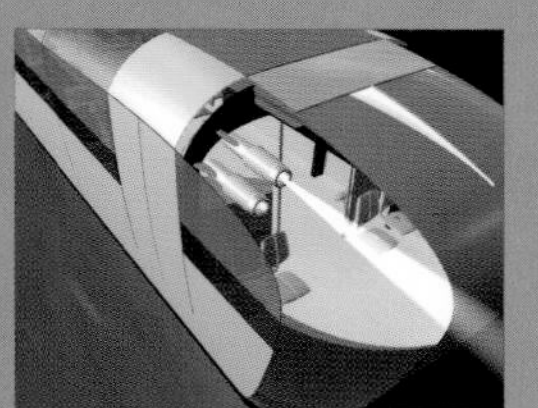

--- fahrerloses airportshuttle; die führerlose bahn "sputnik" verbindet ballungsräume mit ausgelagerten flughäfen -- exterieur: freier raum im kopf des zuges auf grund des fahrerlosen betriebes -- interieur: grosszügige sitzanordnung (viel fussraum und platz für handgepäck) -- informationsdisplays mit abflugzeiten und gateangaben -- zentrales gepäckfach für reisegepäck (cartridgesystem) ---

--- driverless airport shuttle: "sputnik", the driverless railway, connects urban areas with airports in outlying areas -- exterior: free space at the head of the train because of the driverless operation -- generously dimensioned seating arrangements (plenty of leg space and room for hand luggage) -- information displays with airline departure times and gate information -- central luggage compartment (cartridge system) ---

FH JOANNEUM
FACHHOCHSCHULSTUDIENGÄNGE
DER TECHNIKUM JOANNEUM GMBH

2000 (15)

projekt: » bewegung auf schnee und eis «
4. semester/sommer 2000

betreuer: kurt h i l g a r t h (fancy form) - gastprofessor
georg w a g n e r - professor für engineering
werner k l e i s s n e r - walter l a c h - modellbau

thema: der rahmen für diese aufgabenstellung ist wiederum bewusst sehr weit gesteckt: die einzige verbindliche bedingung bei dieser themenstellung ist: "man muss dieses gerät zur fortbewegung auf schnee und/oder eis verwenden können". es geht also um die entwicklung von innovativen winter(sport)geräten, die universell einsetzbar und leicht transportierbar sein müssen. zielgruppen: schnee-, eis- und tourengeher.

project: » locomotion on snow and ice «
4th semester/spring 2000

advisors: kurt h i l g a r t h (fancy form) - guest professor
georg w a g n e r - professor of engineering
werner k l e i s s n e r - walter l a c h - model building

subject: the scope for this problem definition was purposely very broadly-based. the only mandatory requirement was that "one must be able to use this device for locomotion on snow and/or ice". here, we are dealing with the development of innovative winter sport devices, which are universally deployable and easily transportable. the target groups are snow and ice hikers and cross-country hikers.

bewegung auf

sign

p r o j e k t a r b e i t e n d e s f h - s t u d i e n g a n g e s i n d u s t r i a l d e s i g n - g r a z
p r o j e c t s o f t h e f h - d e g r e e c o u r s e i n d u s t r i a l d e s i g n - g r a z

2000
(15)

schnee + eis

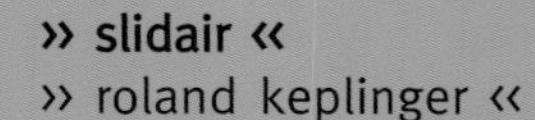

» slidair «

» roland keplinger «

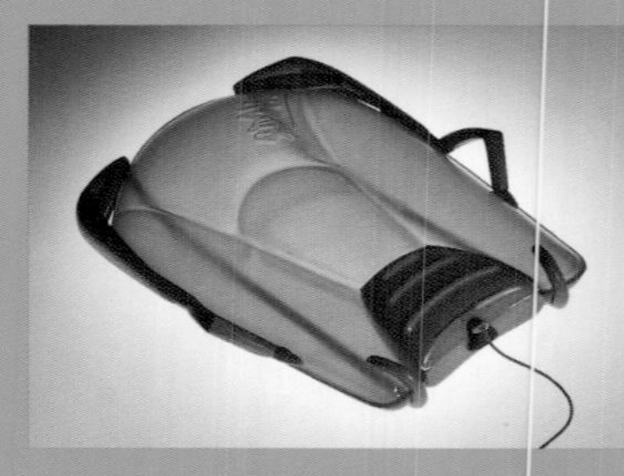

--- "slidair" ist quasi die rasante luftmatratze für den winter -- per integriertem blasebalg leicht und schnell aufzublasen, leicht transportierbar -- durch kufen und griffe ist auch die kurvenfahrt möglich ---

--- "slidair" is the fast air mattress for the winter, as it were -- can be inflated quickly and easily with the built-in pump, easily transportable -- runners and grips make negotiating curves possible ---

» northern star «

» jan mathis «

--- expeditionsschlitten für extremeinsätze in polargebieten -- über einen bügel kann die abdeckplane zu einem biwakzelt aufgespannt werden -- beim aufklappen des bügels wird der schlitten mittels zweier krallen im boden verankert -- der vordere untere teil ist abnehmbar und kann als schalenrucksack verwendet werden ---

--- expedition sled for extreme operations in arctic regions -- the canvas top can be stretched over a bar to form a bivouac tent -- the bar can be folded out to anchor the sled in the ground with two claws -- the front lower part can be removed and used as a hard-sided backpack ---

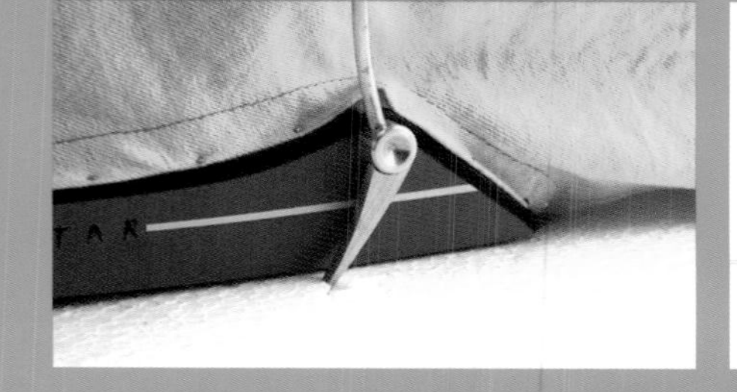

» floats «
» peter körbler «

--- "aggressiv-skaten" bezeichnet eine stilform im funsport -- die praktiker dieses stils sind immer auf der suche nach dem innovativen trick -- dazu zählen v.a. auch manöver wie das rutschen über geländer, handläufe, betonkanten etc. (= grinden) in speziellen funparks ---

--- "aggressive skating" is a skating style in leisure sports -- the practicians of this style are always looking out for innovative tricks -- this includes maneuvers such as sliding over railings, guard rails, concrete ledges, etc. (called "grinding") in special amusement parks ---

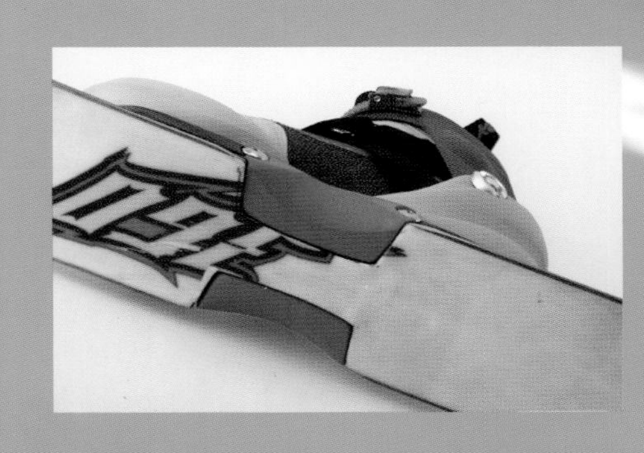

--- "s'nomad" ist ein schneeschuh für snowboarder, der bei abfahrten einfach über ein step-in-bindungssystem mit aufs snowboard geschnallt werden kann -- vorteil: extrem schneller wechsel zwischen schneeschuh- und snowboardeinsatz ---

--- "s'nomad", a snow shoe for snowboarders, has a step-in binding system and can simply be buckled onto the snowboard for the down-hill descent -- advantage: extremely quick change between snow shoe and snowboard use ---

» s'nomad «
» martin schnitzer «

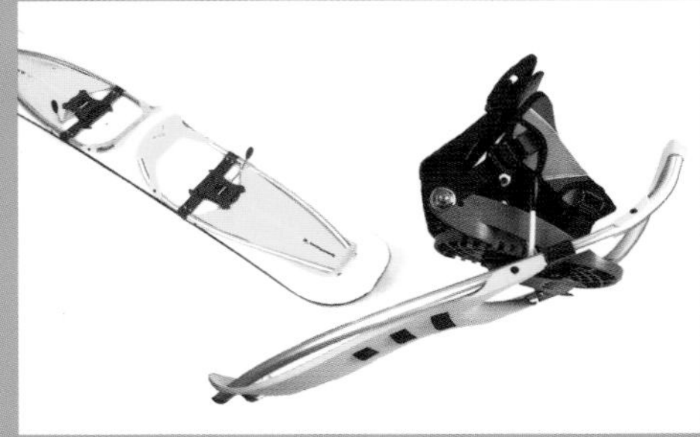

--- es handelt sich hier um einen rucksack mit rettungsschaufel, der es möglich macht, einen verletzten aus einem gefahrengebiet zu bergen (der hubschrauber kann nicht überall landen) -- der verletzte wird sitzend auf der rettungsschaufel abtransportiert ---

--- this is a backpack with a kind of rescue shovel that makes it possible to evacuate an injured person from a dangerous area (helicopters cannot land everywhere) -- the injured person is transported in a sitting position on the rescue shovel ---

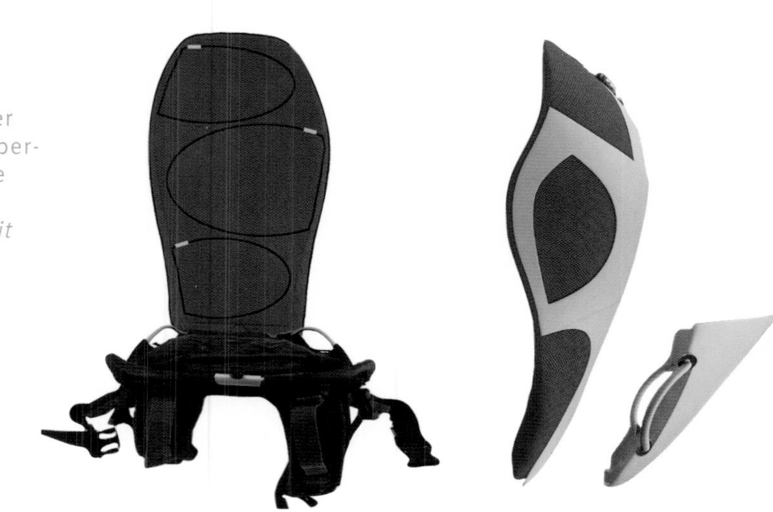

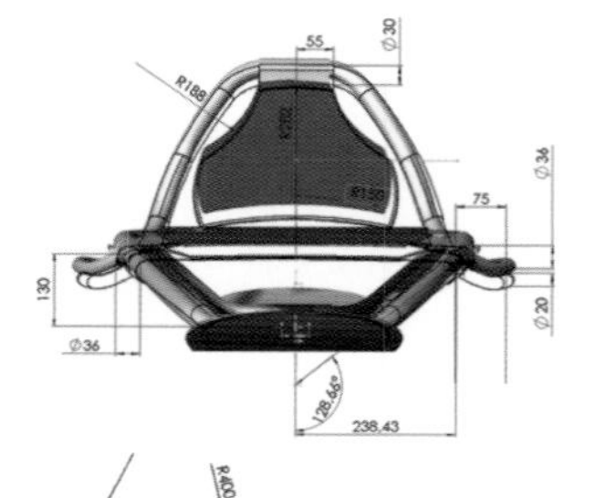

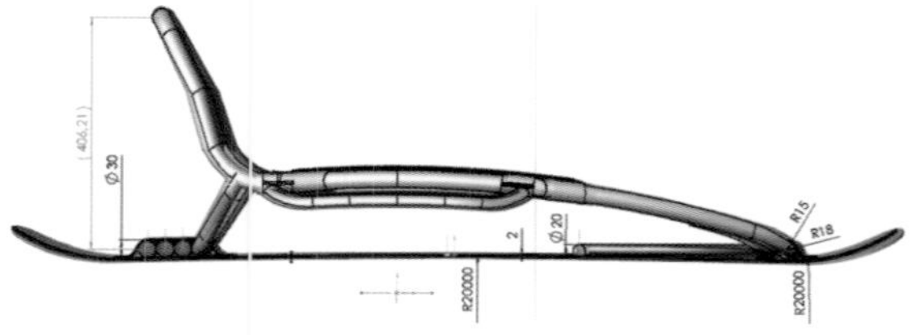

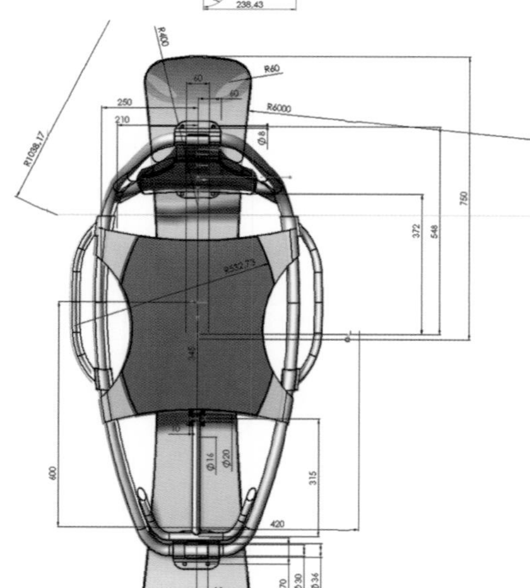

--- die grundidee von "sonic frame" ist: - je tiefer man sich über dem boden fortbewegt, desto höher ist die empfindung der dabei erreichten geschwindigkeit -- ziel: ein board mit dem es möglich ist, die piste sitzend hinunterzusausen ---

--- the basic idea behind "sonic frame" is: - when moving, the closer one is to the ground, the faster one seems to be going -- goal: a board that makes it possible to zip down the trail in a sitting position ---

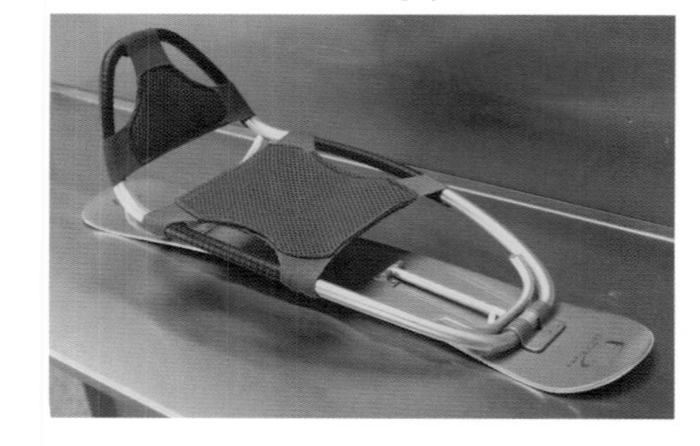

» sonic frame «

» rainer trummer «

» tortuga «
» elger oberwelz «

» ray - pack «
» robert hitthaler «

--- zielgruppe für den "raypack" sind all jene, die unabhängig von der aufstiegsart pisten, waldwege oder abhänge in einer, dem rodeln sehr ähnlichen, aber doch neuen art "downhillen" wollen -- aber auch snowboarder und skifahrer, die die abwechslung lieben und den "raypack" einfach dabei haben ---

--- the target group for the "raypack" are those who, independent of the manner of ascent, want a new way to race downhill on trails, forest trails or mountain slopes that is very similar to sledding -- this is also something for snowboarders and skiers who like variety and simply want to have the "raypack" with them ---

FH JOANNEUM FACHHOCHSCHULSTUDIENGÄNGE DER TECHNIKUM JOANNEUM GMBH

2000 (16)

projekt: » leichtfahrzeuge für indien «
– in kooperation mit steyr india
6. semester/sommer 2000

betreuer: gerhard h e u f l e r - gerald k i s k a - professoren für design -
georg w a g n e r - professor für engineering -
werner k l e i s s n e r - walter l a c h - modellbau –

thema: in vielen ländern des fernen ostens hält sich ein fahrzeug, das nie dafür konzipiert worden war: der motorroller. meist zweckentfremdet zum transport ganzer familien, tieren oder schwerster lasten, besticht er durch günstigen preis, leichte reparierbarkeit und simplen aufbau. es gilt also fahrzeuge zu konzipieren, die auf der bekannten rollerbauweise aufbauen (triebsatzschwinge), aber speziell auf die besonderen einsatzzwecke dieser länder abgestimmt sind.

project: » light vehicles for india «
- in collaboration with steyr india
6th semster/spring 2000

advisors: *gerhard h e u f l e r - gerald k i s k a - professors of design -*
georg w a g n e r - professor of engineering -
werner k l e i s s n e r - walter l a c h - model building -

subject: *in many countries of the far east, a vehicle, which was not originally designed for it, remains popular - the motorscooter. it is usually misappropriated to transport whole families, animals, or very heavy loads, since its low cost, simple construction, and the fact that it is easy to repair make it attractive for the consumer. therefore, vehicles need to be designed, which use the construction of the motorscooter as a jumping-off point (shifting arm), but are specially coordinated to fit the type of use in these countries.*

leichtfahrzeuge

sign

projektarbeiten des fh-studienganges industrial design-graz

projects of the fh-degree course industrial design-graz

2000
(16)

für indien

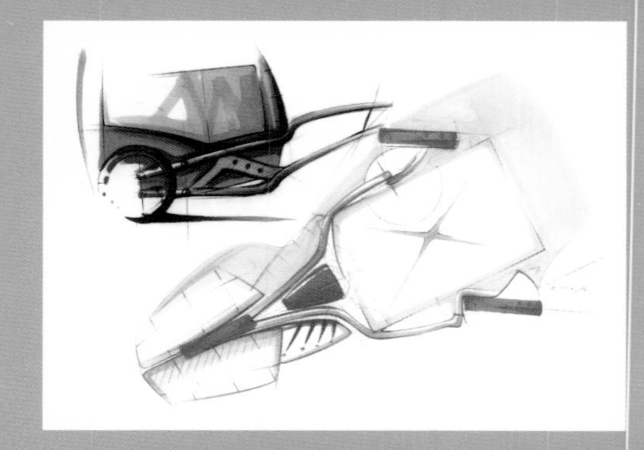

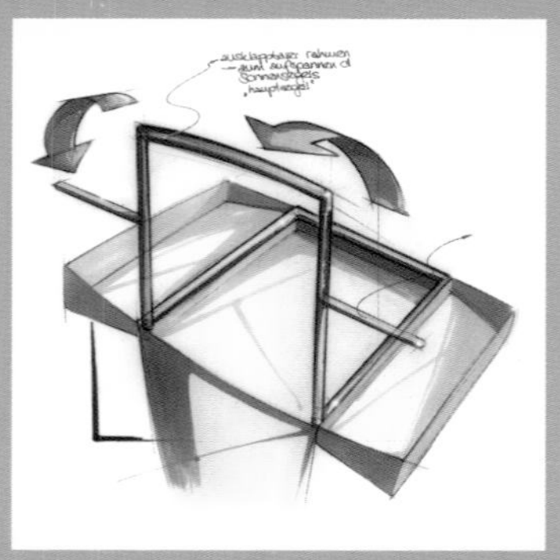

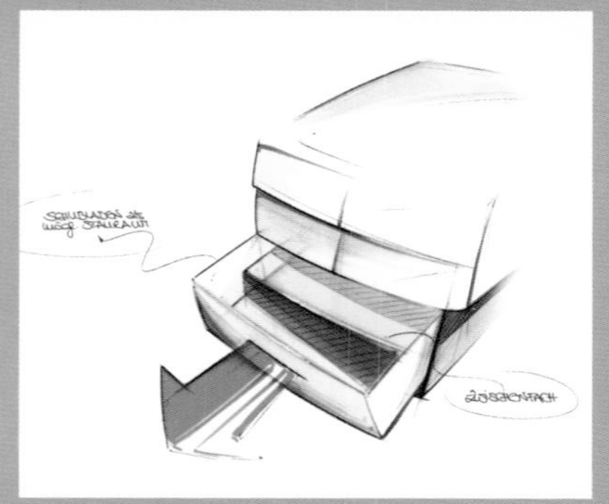

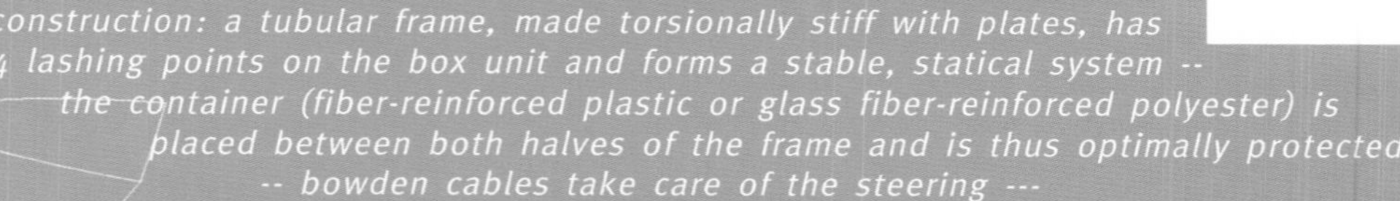

--- konstruktion: ein rohrrahmen, der mit platten torsionssteif wird und mit vier angriffspunkten an der box ein stabiles statisches system bildet -- der container (faserverstärkter kunststoff oder glasfaserverstärkter polyester) ist zwischen den beiden rahmenhälften platziert und so optimal geschützt -- die lenkung erfolgt über bowdenzüge ---

--- construction: a tubular frame, made torsionally stiff with plates, has 4 lashing points on the box unit and forms a stable, statical system -- the container (fiber-reinforced plastic or glass fiber-reinforced polyester) is placed between both halves of the frame and is thus optimally protected -- bowden cables take care of the steering ---

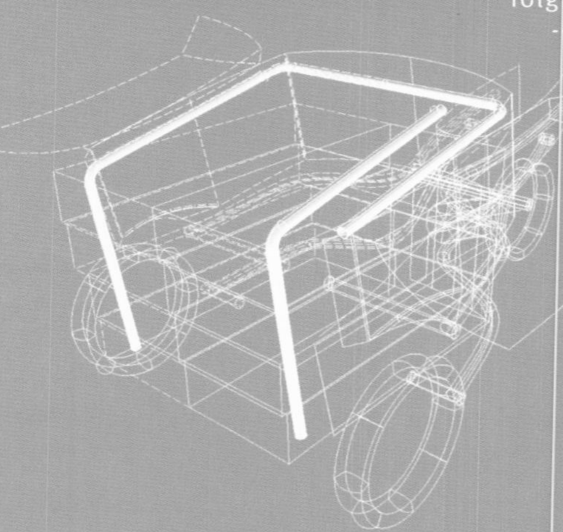

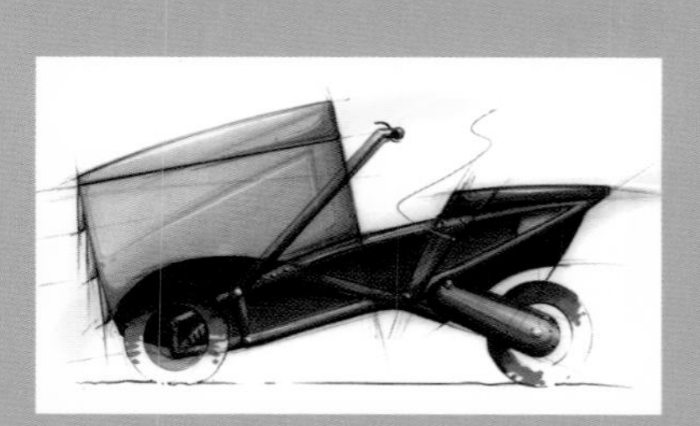

» tripod «

» beate dörflinger, marc ischepp «

--- die zielgruppe, die wir mit unserem mobilen verkaufsstand ansprechen wollen, sind kleinhändler oder sogar anbieter von snacks und imbisshappen -- dieses dreirad ist flexibel, weil es ein stapelsystem per schubladen anbietet und auch in windeseile zu einem kleintransporter für grössere waren umgebaut werden kann ---

--- the target group for our mobile kiosk are tradespeople or providers of lunch food and snacks -- the kiosk has three wheels and is flexible because its batch system with modules that can be converted quickly to a small transporter for larger loads ---

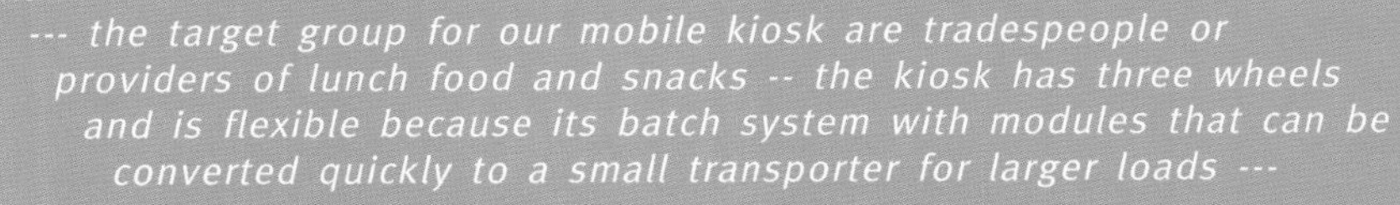

--- funktion im einsatz: - lastentransporter -- zielgruppe: - gewerbetreibende, - händler, - handwerker, - transporteure, - botendienste -- hauptmerkmale: - wendigkeit, - geringe breite, - grosse ladefläche, - 360° drehende antriebseinheit (ermöglicht minimalen wendekreis, hohe wendigkeit und rückwärtsfahrt ohne retourgang) ---

--- load transporter -- target group: - tradespeople, - merchants, - artisans, - haulers, - delivery people -- main characteristics: - maneuverability, - narrow width, - large loading surface, - propulsion unit can be swiveled 360° (makes a minimal turning circle, extreme maneuverability and backing up without a reverse gear possible) ---

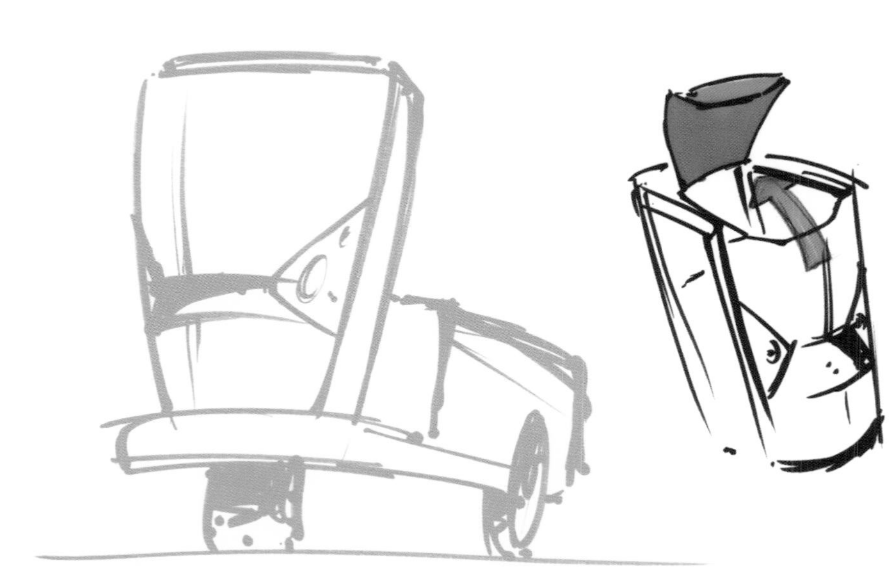

» drehmoment *torque* «

» daniel dockner, gerald krenn, josef lintl «

heck zum beladen abgesenkt *the tail end is lowered for loading*

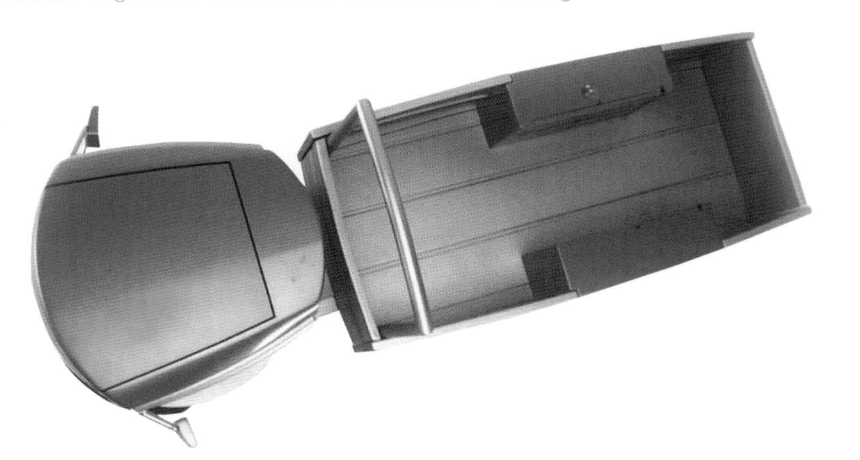

» scooter «

» albert ebenbichler, gertraud körner, karin krumphals «

› › › ›

› › › ›

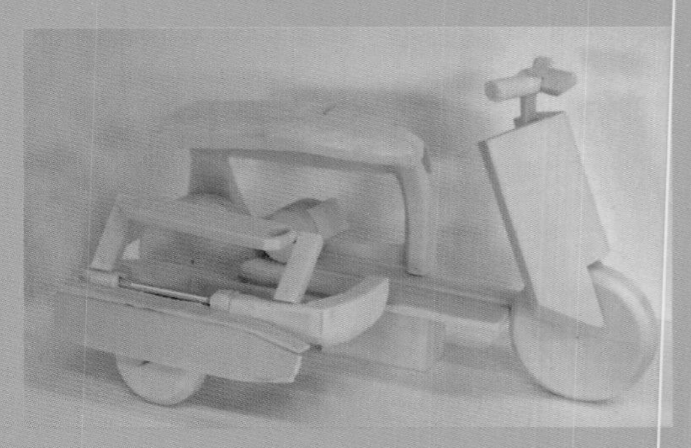

--- idee: eine ausklappbare sitzgelegenheit und transportfläche in das volumen eines "scooters" zu integrieren -- ein klappsystem (3-rad) ermöglicht den zusätzlichen transport von gütern oder zweier weiterer personen ---

--- idea: to integrate a foldable seat and loading surface into the entirety of a "scooter" -- a foldable system (3 wheels) makes it possible to transport goods and to fit in two more passengers ---

» sammeltaxi für indien ***collective taxi for india*** **«**

» martin dick, peter kalsberger «

--- das taxi ist für 7 personen ausgelegt (6 passagiere, 1 fahrer) -- der fahrzeugrahmen ist als trägerrohrrahmen konzipiert -- hinten antriebseinheit (motor und triebsatzschwinge) -- quer zur fahrtrichtung positionierte sitze, die seitlich am mittelträger montiert sind, gewährleisten schnelles ein- und aussteigen -- ausserdem sinken dadurch die fahrzeugausmasse auf ein minimum ---

--- the taxi is designed for 7 persons (6 passengers, 1 driver) -- the vehicle frame is designed as a tubular support frame -- propulsion unit in the rear (engine and propelling arm) -- seats, mounted on the side of the central support and positioned at right angles to the traveling direction, make quick getting on and off possible -- vehicle dimensions are minimal ---

» drag «

» gerfried gaulhofer, martin mathy «

--- überbevölkerung und die nicht gewährleistete umverteilung der lebensnotwendigen güter lassen personen, die am rande des versorgungsnetzes leben, ohne die chance auf eine zukunft -- in den regenzeiten kann die versorgung von kleineren dörfern abseits der hauptadern nicht gewährleistet werden -- das all-terrain-vehicle "drag" ermöglicht diesen brückenschlag ---

--- overpopulation and difficulty in redistribution of vital goods leave people, who live beyond the reaches of the supply network without opportunities for their future -- during the monsoons, keeping the smaller villages off the beaten track supplied is difficult -- the all-terrain vehicle "drag" bridges this gap ---

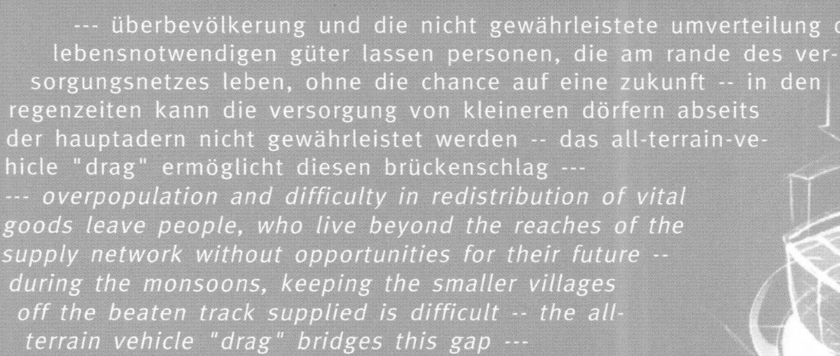

projekt: » diplomarbeiten « – mit div. kooperationspartnern
8. semester/sommer 200

betreuer: gerhard h e u f l e r - gerald k i s k a - professoren für design -
georg w a g n e r - professor für engineering -
werner k l e i s s n e r - walter l a c h - modellbau –

thema: drei themenbereiche bilden die schwerpunkte des zweiten diplomjahrganges: fahrzeuge, elektronische geräte und produkte für die freizeit. die palette reicht vom motorrad über pkws bis zum wasserfahrzeug, vom digitalen collegeblock über kommunikationssysteme bis zum snowboardschuh. als kooperationspartner konnten audi, design a storz, fancy form, ktm, steyr fahrzeugtechnik, philips und die forschungsgesellschaft joanneum research gewonnen werden.

project: » theses « – with different collaborators
8th semester/spring 2000

advisors: *gerhard h e u f l e r - gerald k i s k a - professors of design -*
georg w a g n e r - professor of engineering -
werner k l e i s s n e r - walter l a c h - model building -

subject: *three fields are the focal points of the second year of graduates: vehicles, electronic devices, and leisure-time products. the range goes from the motorcycle to passenger cars to water vehicles, from the digital writing pad to communications systems to snowboarding shoes. the following companies have been enlisted as collaboration partners: audi, design a storz, fancy form, ktm, steyr fahrzeugtechnik, philips, and joanneum research.*

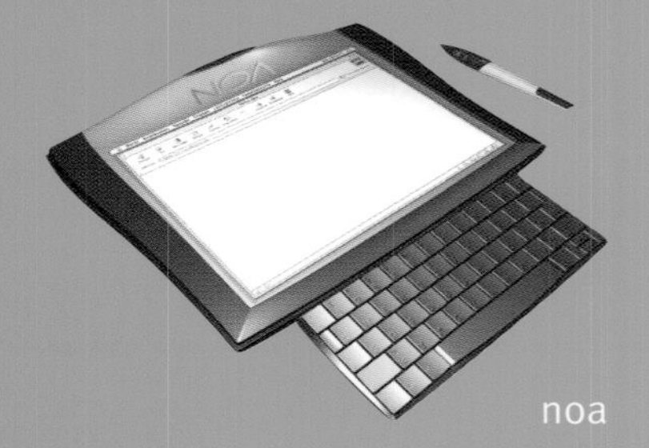
noa

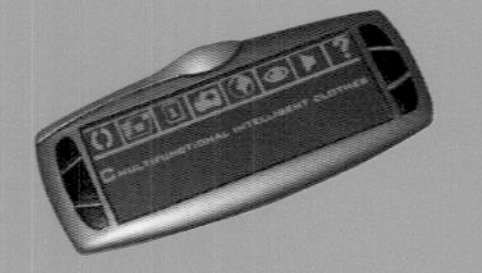
mic

transvision

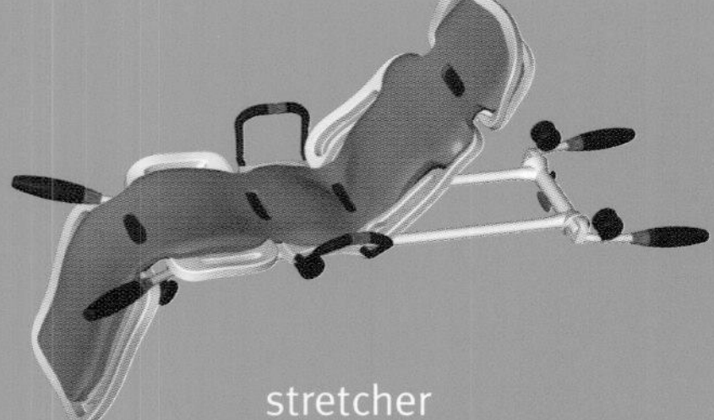
stretcher

diplom-arbeit

sign

projektarbeiten des fh-studienganges industrial design-graz
projects of the fh-degree course industrial design-graz

2000
(17)

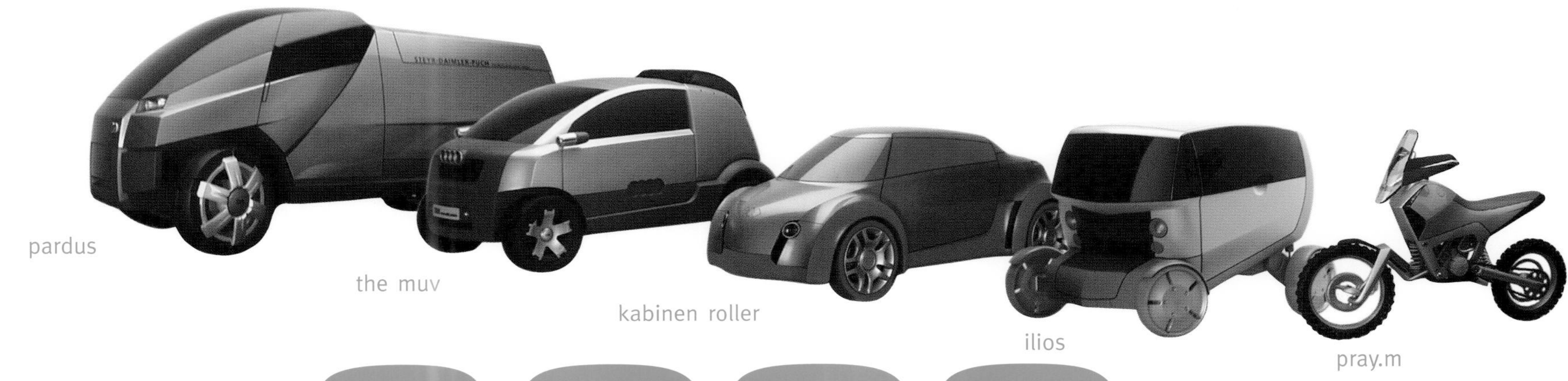

en 2000

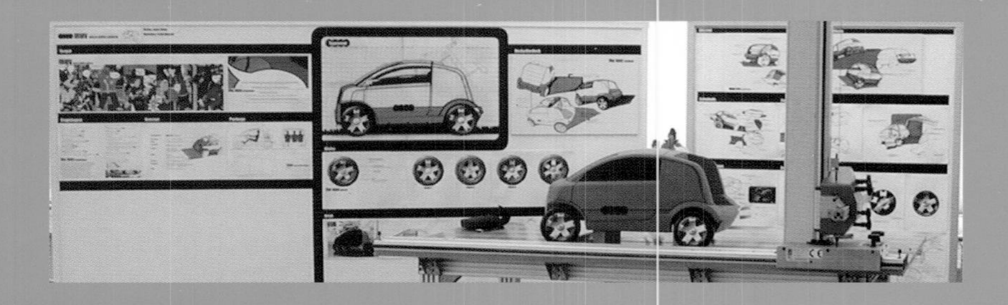

--- "muv" steht für micro utility vehicle und ist in kooperation mit audi design entstanden -- er ist drei meter kurz, aber gleich breit wie eine limousine -- der antrieb erfolgt über eine brennstoffzelle und zwei elektromotoren -- die fahrerkabine bietet platz für drei nebeneinander angeordnete personen -- der variable laderaum bietet platz für snowboards, surfbretter, drei mountainbikes oder einen rucksack ---

--- "muv", which stands for micro utility vehicle, was designed in collaboration with audi design -- it is 3 meters short, but as wide as a sedan -- a fuel cell and 2 electric engines take care of the propulsion -- the driver's compartment has room for three persons seated next to one another -- the variable loading area has room for snowboards, surfboards, 3 mountain bikes or a backpack ---

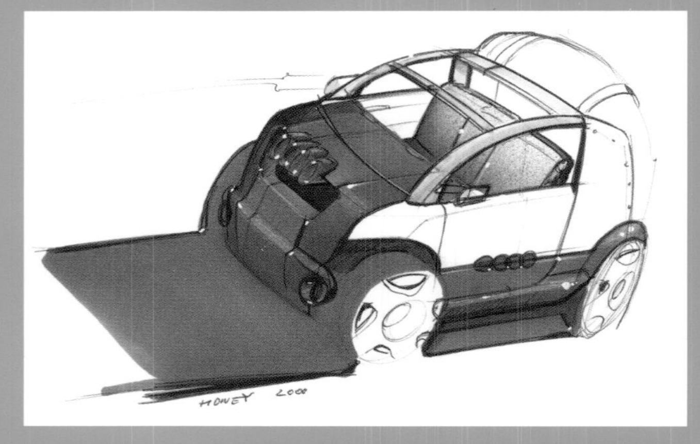

» the muv «
» julian hönig «

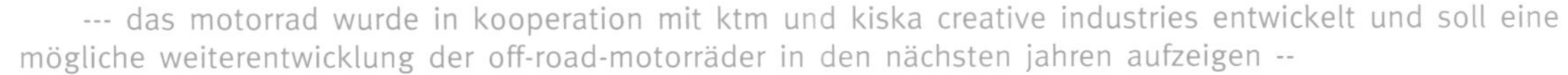

--- das motorrad wurde in kooperation mit ktm und kiska creative industries entwickelt und soll eine mögliche weiterentwicklung der off-road-motorräder in den nächsten jahren aufzeigen -- merkmale: - 1. der durchfliessende luftstrom, der eine verbesserung der aerodynamik bringt und grössere kühlflächen zulässt, - 2. das konstruktive rückgrat, das die schnittstelle zwischen fahrer und fahrwerk deutlich hervorhebt und gleichzeitig den korpus zur aufnahme der technik bildet, - 3. der hydraulikmotor in der vordernabe erhöht die geländetauglichkeit ---

--- the motorcycle was developed in collaboration with ktm and kiska creative industries and will possibly be an evolutionary development of off-road motorcycles in the coming years -- characteristics: - 1. an air current flows through it and brings an improvement of the aerodynamics and allows larger cooling surfaces, - 2. the constructional backbone that clearly highlights the interface between driver and chassis and at the same time creates the carcass that will accommodate the technology, - 3. the hydraulic engine in the front hub increases its suitability for off-road use ---

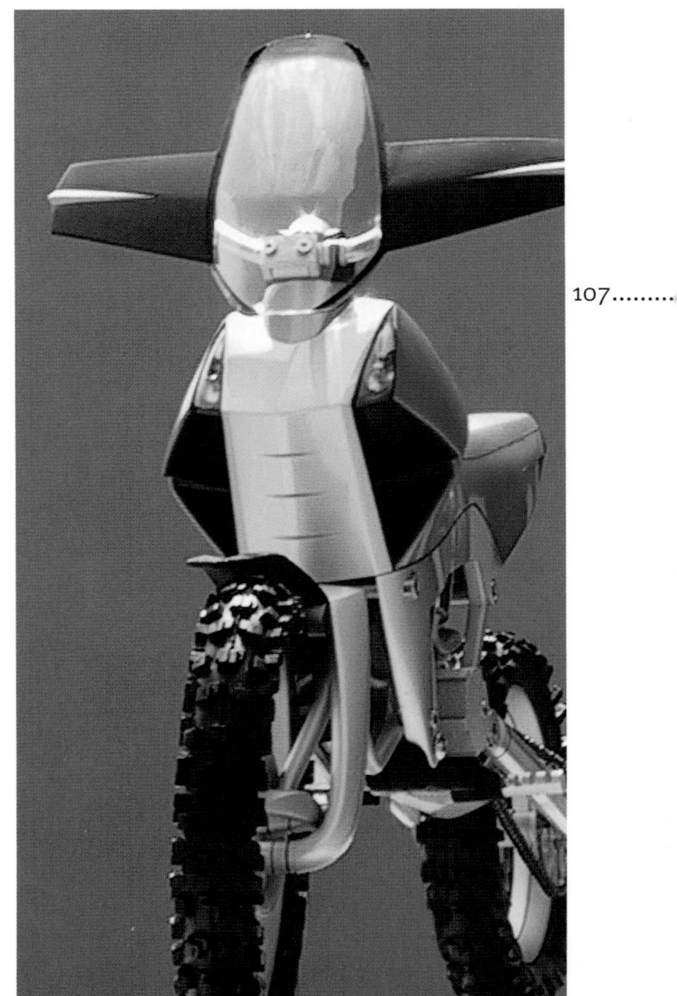

» pray.m «
» christoph gredler «

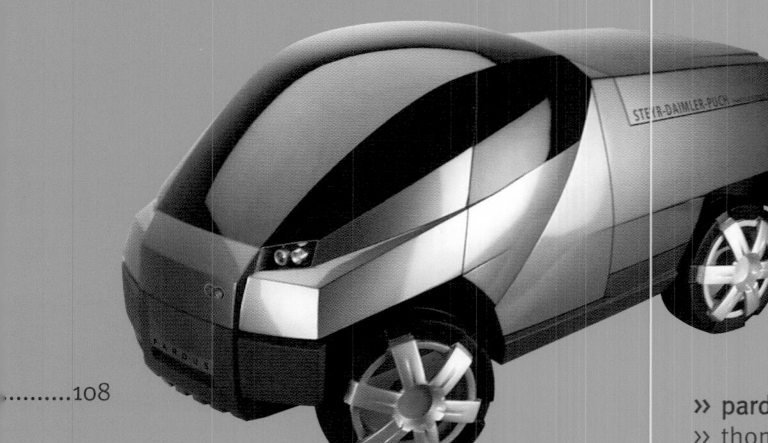

--- "pardus" ist eine studie für ein geländegängiges nutzfahrzeug für den einsatz von feuerwehr, rettung, polizei, aber auch für den privaten gebrauch -- die arbeit ist in kooperation mit dem fahrzeughersteller steyr-daimler-puch entstanden ---

--- "pardus" is a study for a cross-country utility vehicle for fire brigade, ambulance, and police use, but also for personal use -- this study was made in collaboration with the vehicle manufacturer steyr-daimler-puch ---

» pardus «

» thomas binder «

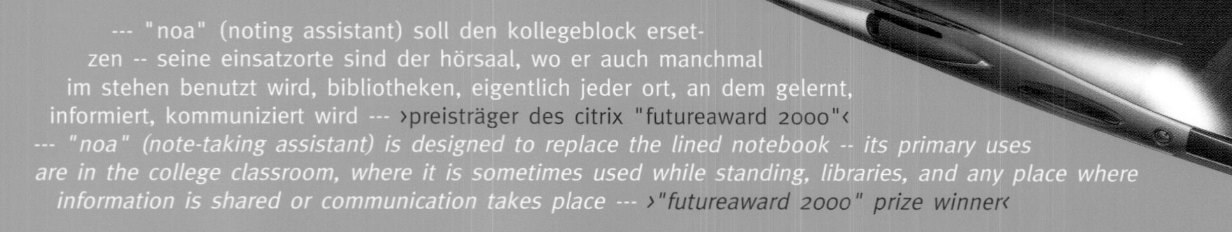

--- "noa" (noting assistant) soll den kollegeblock ersetzen -- seine einsatzorte sind der hörsaal, wo er auch manchmal im stehen benutzt wird, bibliotheken, eigentlich jeder ort, an dem gelernt, informiert, kommuniziert wird --- ›preisträger des citrix "futureaward 2000"‹

--- "noa" (note-taking assistant) is designed to replace the lined notebook -- its primary uses are in the college classroom, where it is sometimes used while standing, libraries, and any place where information is shared or communication takes place --- ›"futureaward 2000" prize winner‹

» noa «

» petrus gartler «

›› **ilios** ‹‹
›› christian becker ‹‹

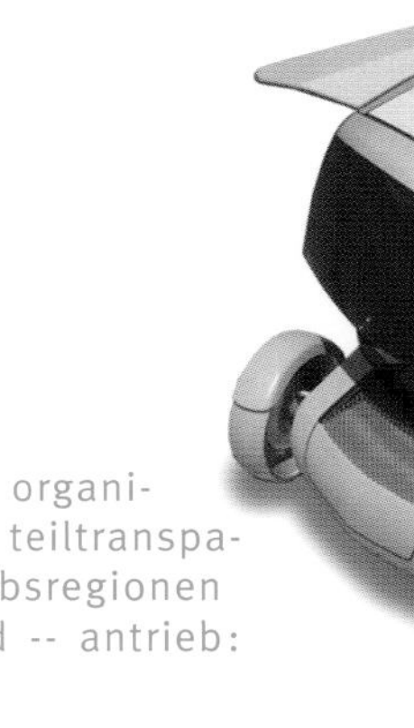

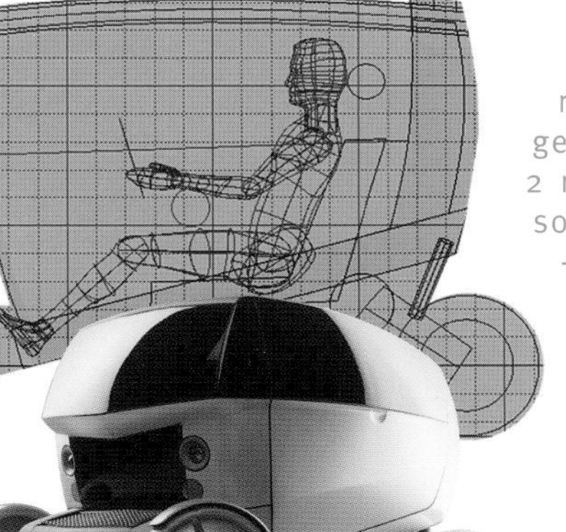

--- das solarmobil "ilios" nutzt die möglichkeiten organischer solarzellen: - aufbringung auf freiformflächen, - teiltransparenz -- es ist als mietfahrzeug in warmen, südlichen urlaubsregionen gedacht, in denen nur emissionsfreie fahrzeuge erlaubt sind -- antrieb: 2 radnabenmotoren hinten, akkus unter den sitzen -- solartechnik: joanneum research ---

--- the solar energy automobile "ilios" uses organic solar cells: application on free-form surfaces makes partial transparency possible -- it is designed to be a rental vehicle in warm vacation spots in southern regions, where only emission-free vehicles are permitted -- propulsion: 2 wheel hub engines in the rear, rechargeable batteries under the seats -- solar technology: joanneum research ---

--- gestaltung eines zweisitzigen kleinstautomobils -- formensprache des messerschmittkabinenrollers tg 500 -- konsequenter leichtbau erlaubt trotz schwacher motorisierung gute beschleunigung -- sportliches auftreten, ohne aggressive gestaltungselemente -- kooperation mit design a storz ---

--- design of a tiny two-seater automobile -- form of the messerschmitt cabin roller tg 500 -- consistent lightweight design allows good acceleration despite a weak engine -- sporty appearance without aggressive design element -- collaboration with design a storz ---

›› **kabinen roller** ‹‹
›› leonard natterer ‹‹

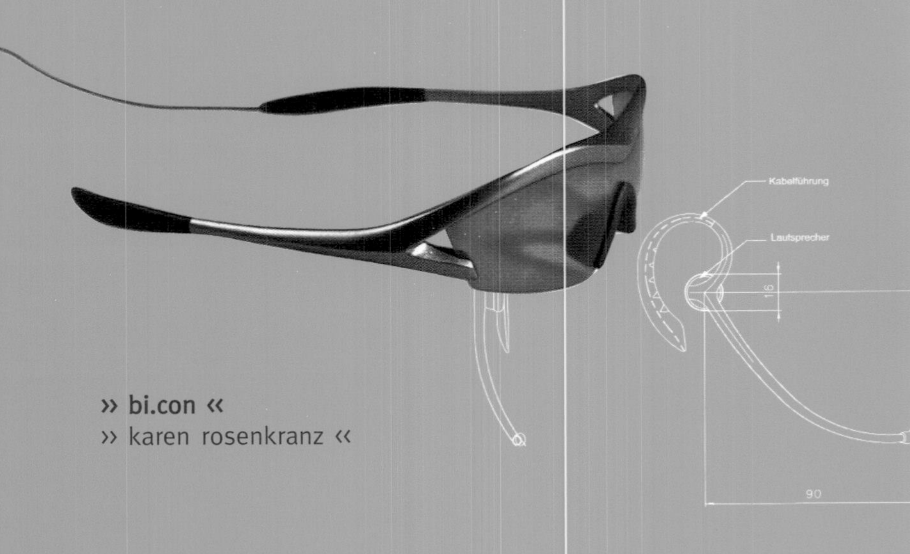

» bi.con «
» karen rosenkranz «

--- innovatives kommunikationssystem für fahrradkuriere -- die brille ermöglicht die darstellung eines virtuellen stadtplans oder die anzeige einer auftragsliste -- bedienung über voice control -- das sourcepad schickt die auftragsbestätigung automatisch an die zentrale --- ›preisträgerin des citrix "futureaward 2000"‹

--- innovative communications system for bicycle couriers -- the eyeglasses make it possible to show a virtual city map or a job list -- operated by voice control -- the source pad sends the job confirmation automatically to the dispatcher --- ›"futureaward 2000" prize winner‹

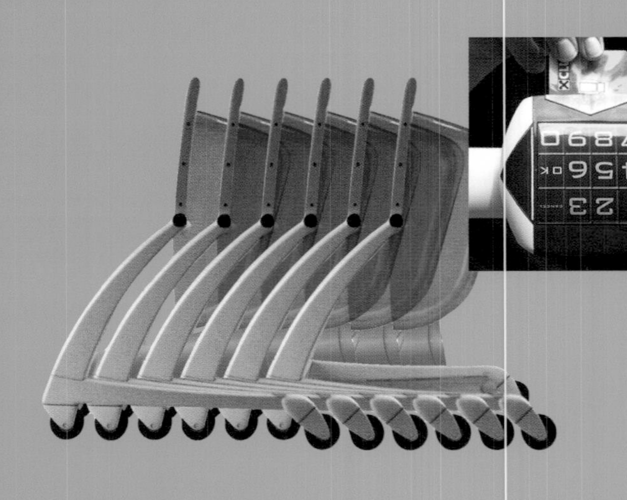

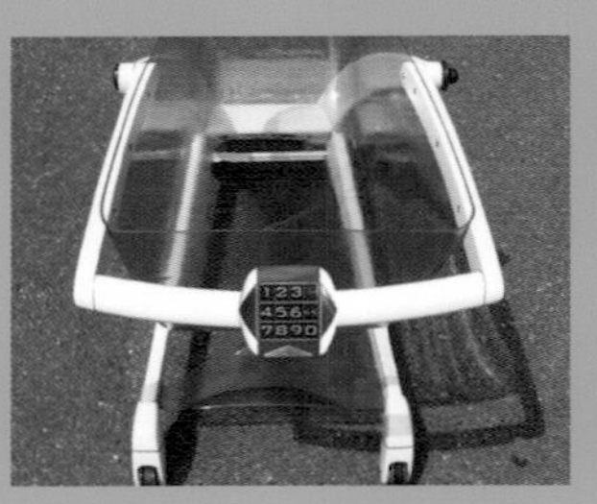

» intelli cart «
» rudolf steiner «

--- die grundidee dieses konzeptes ist es, den einkaufsablauf für den konsumenten komfortabler zu gestalten -- die info-box gibt auskunft über waren, preise, angebote und hilft dem konsumenten, bereiche schneller zu finden -- die abrechnung an den kassen erfolgt über ein funk-system ---

--- the basic concept of this project is to make shopping more consumer-friendly -- the info box gives information about merchandise, prices, and specials and helps the consumer to find what he/she is looking for more quickly -- paying at the cash register takes place via a radio system ---

» mic «
» barbara keimel «

--- "mic" = multifunctionel intelligent clothes -- der stoff ist temperaturregulierend und mit solarzellen ausgerüstet -- der multimediaterminal ist im ärmel integriert; er beinhaltet mobiltelefon, internet, gps und digitale kamera -- solartechnik: joanneum research ---

--- "mic" = multifunctional intelligent clothes -- the fabric is equipped with solar cells and regulates the temperature -- the multimedia terminal is integrated into one of the sleeves and contains a cell phone, internet connection, gps, and a digital camera -- solar technology: joanneum research ---

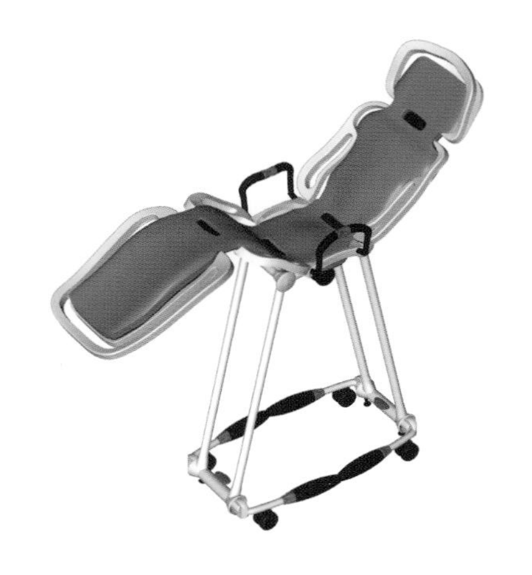

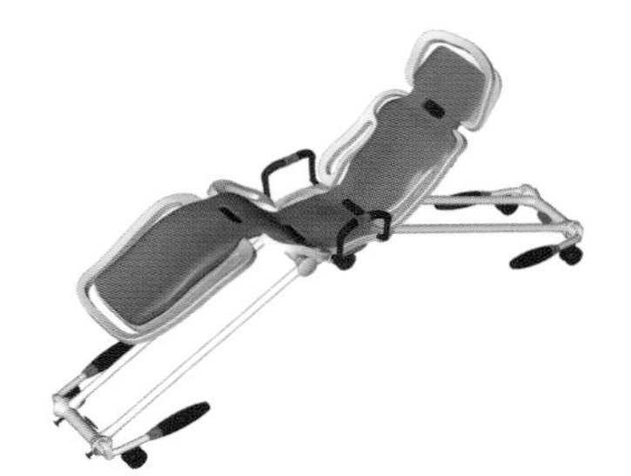

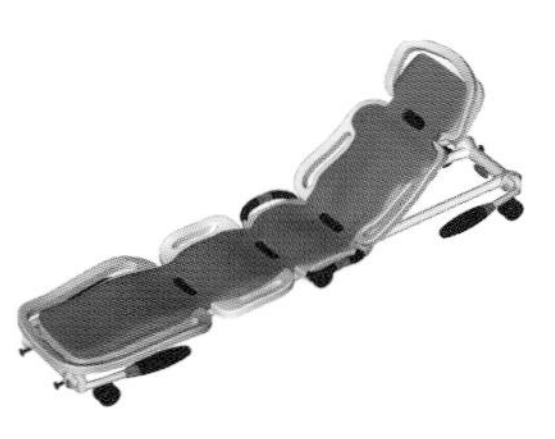

» stretcher «
» karin ruthardt «

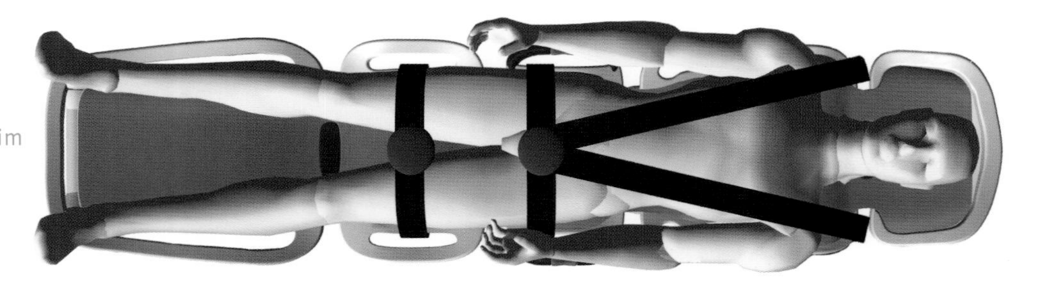

--- "stretcher" wurde für den transport von verletzten oder schwerkranken im flugzeug oder helikopter konzipiert -- das erscheinungsbild basiert auf der optischen trennung zwischen dem organisch ausgeformten liegeteil und dem funktionalen, technischen rahmen ---

--- "stretcher" was designed for the transport of injured or severely ill persons in airplanes or helicopters -- the main characteristic is the optical separation of the organically formed cot and the functional, technological framework ---

--- audiovisuelle unterhaltungseinheit -- dvd, audio-cd und cdi sowie videobeamersystem integriert -- zielgruppen: cineasten und game-freaks mit hang zum nomadentum -- in kooperation mit philips design ---

--- audiovisual entertainment unit -- dvd, audio, cd and cdi, as well as video beamer systems are integrated -- target group: film and games aficionados who have a penchant for being nomads -- in collaboration with philips design ---

» transvision «
» christian zwinger «

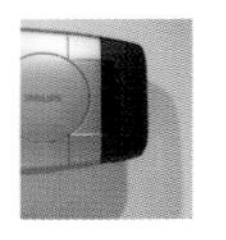

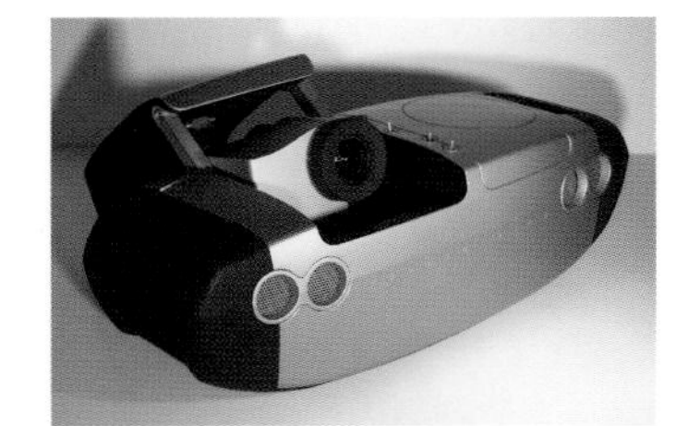

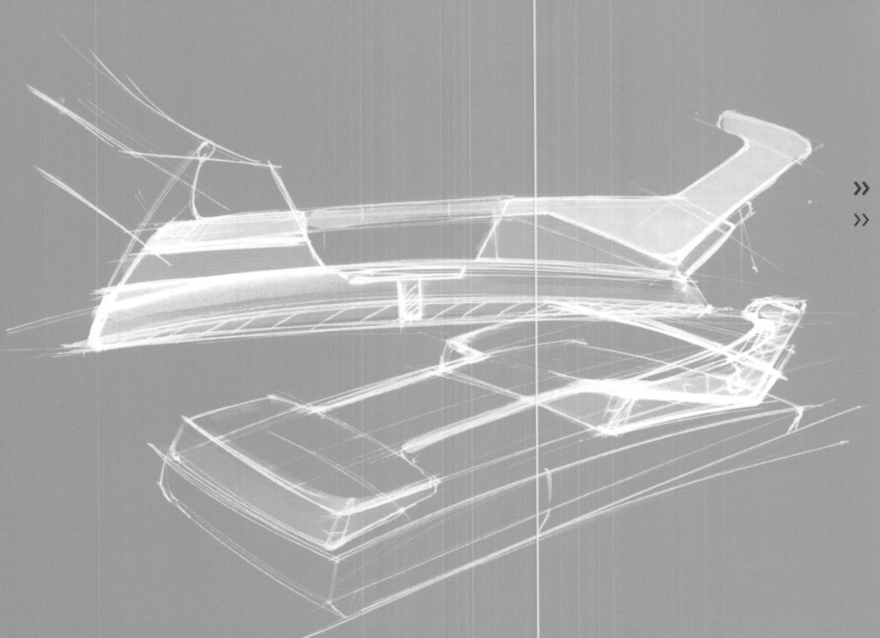

» tri.core «
» volker pflüger «

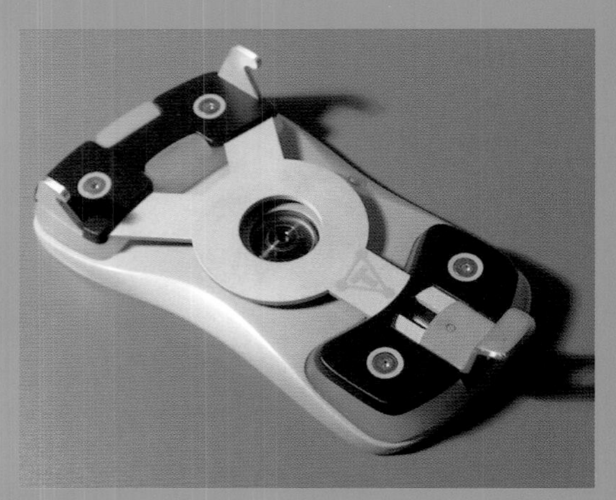

herausnehmbarer innenschuh *removable inner shoe*

kunststoffschale *plastic housing*

textil-softboot *soft boot made of fabric*

vorlageverstellschraube
adjusting screw

klettverschlüsse
velcro fasteners

vorlagendämpfung *damping*

haltegurt
stability strap

zentrales öffnungselement
central opening element

sohle *sole*

aufnahmezone vorne *front anchorage*

--- neuer lösungsansatz für ein komplettes snowboard-step-in-bindungssystem -- zielgruppe: - erfahrene boarder und - experimentierfreudige fahrer -- die hauptidee der bindungsaufnahme besteht in einem neuartigen dreipunktsystem -- dabei wird der vorderfuss an zwei punkten gehalten, der dritte aufnahmepunkt sitzt unter der ferse -- in kooperation mit fancy form ---

--- a new solution concept for a complete snowboard step-in binding system -- target group: - experienced snowboarders and - those who like to experiment -- the central idea of the binding fixture consists of a new kind of three-point system -- the front part of the foot is anchored at two points and the third point is under the heel -- in collaboration with fancy form ---

» m-coach «
» martin wöls «

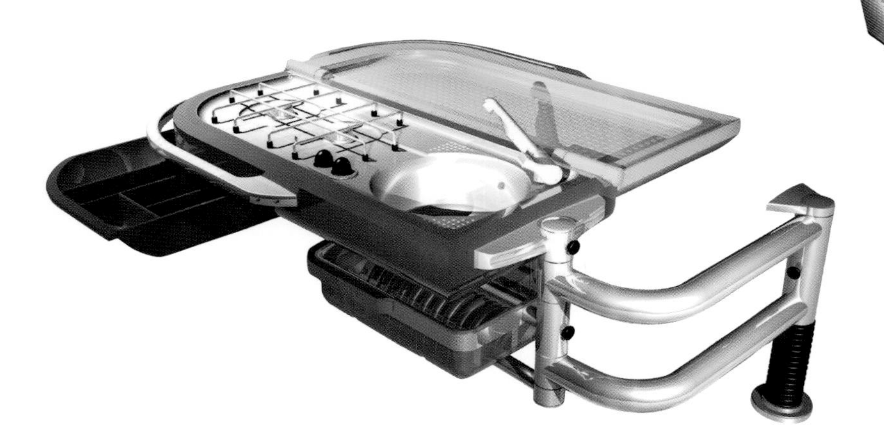

--- "m-coach" ist eine wohnmobilstudie, die auf einem handelsüblichen van aufbaut -- es wurden ein ausklappbares küchenmodul und eine abnehmbare campingbox bearbeitet ---

--- "m-coach" is a trailer study that builds on commercially available vans -- a fold-out kitchen module and a removable camping box were projected ---

» maya «
» valerie trauttmansdorff-weinsberg «

--- durch ausschwenken des bootskörpers wird das solarvehikel zur badeinsel -- nicht "need for speed", sondern "je langsamer, desto deutlicher" ist das motto -- das projekt entstand in kooperation mit joanneum research ---

--- by swinging out the body of the boat, the solar vehicle becomes an island for swimming -- the motto is not the "need for speed", but "the slower, the clearer" -- the project was done in collaboration with joanneum research ---

» emusion «
» roland lindner «

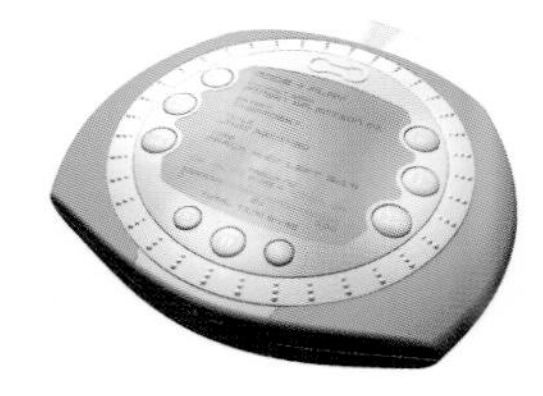

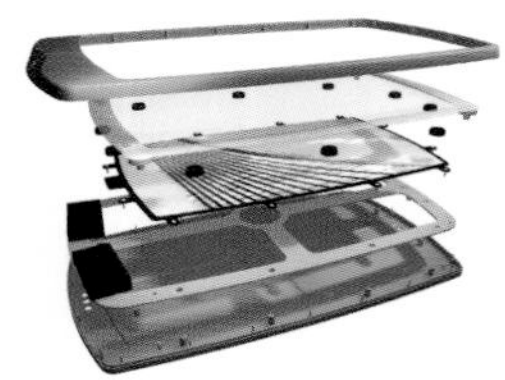

--- die idee war, eine verbindung zwischen ton und bild zu schaffen, die eine neue art der entspannung durch zusätzliche visuelle stimulation bietet ---

--- the idea was to create a connection between sound and image that offers a new kind of relaxation through additional visual stimulation ---

projekt: » shaving and grooming «
- in kooperation mit payer electric shaver
3. semester/winter 2000/01

betreuer: gerhard h e u f l e r - gerald k i s k a - professoren für design
georg w a g n e r - professor für engineering -
werner k l e i s s n e r - walter l a c h - modellbau –

thema: rund um das thema haarpflege (kopfhaar und körperbehaarung) hat sich in den letzten jahren ein bedeutender markt entwickelt und wird in den nächsten jahren noch weiter an bedeutung gewinnen. derzeitige modetrends unterstützen diese entwicklung massiv. unterschiedlichste zielgruppen werden allerdings völlig unterschiedliche anforderungen an produkte in diesem bereich stellen. aufgabe ist es, denkanstösse für zukünftige produktstrategien zu entwickeln.

project: *» shaving and grooming «*
- in collaboration with payer electric shaver
3rd semester/fall 2000/01

advisors: *gerhard h e u f l e r - gerald k i s k a - professors of design -*
georg w a g n e r - professor of engineering -
werner k l e i s s n e r - walter l a c h - model building -

subject: *the subject of hair care (both hair on one's head and body hair) has become a substantial market in the last few years, and it will become even more significant in the coming years. current fashion trends support this development to a great extent. however, very different target groups will have completely different requirements as far as products in this field are concerned. the task here is to develop new ideas for future product strategies.*

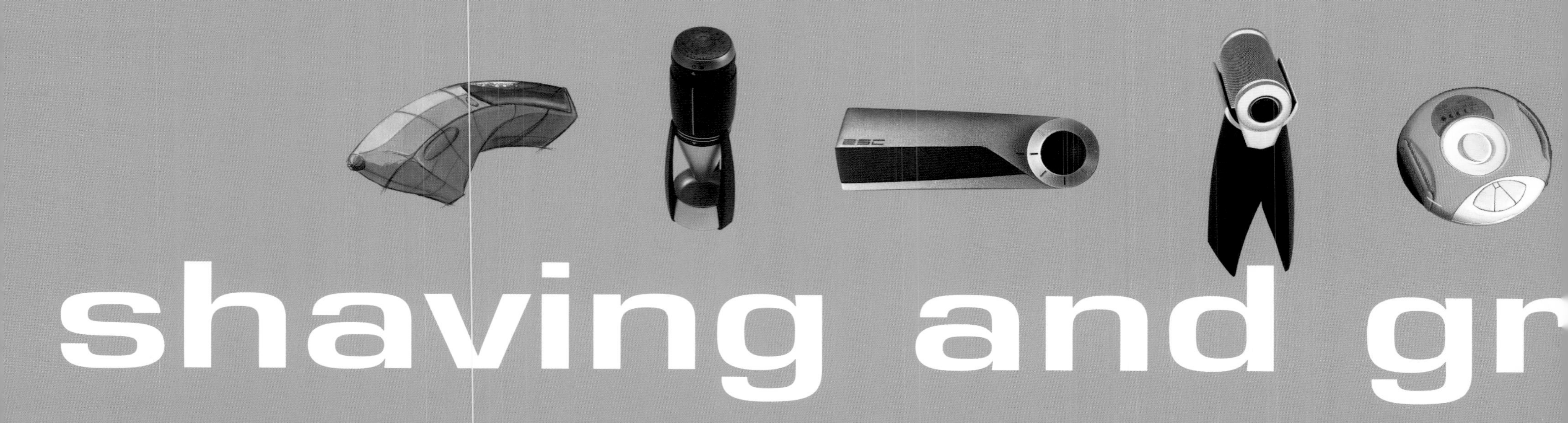

sign

projektarbeiten des fh-studienganges industrial design-graz
projects of the fh-degree course industrial design-graz

2000/01
(18)

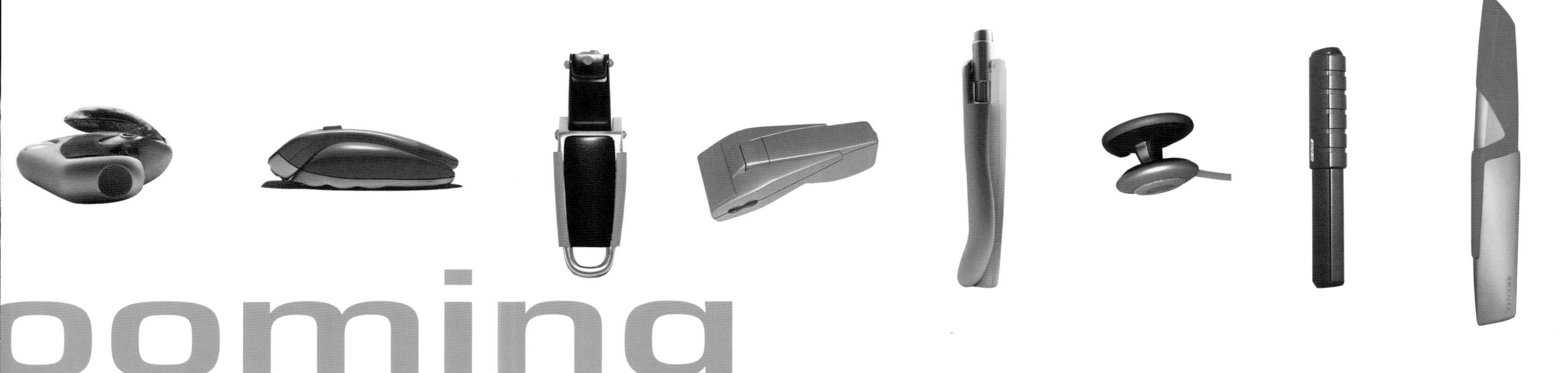

ooming

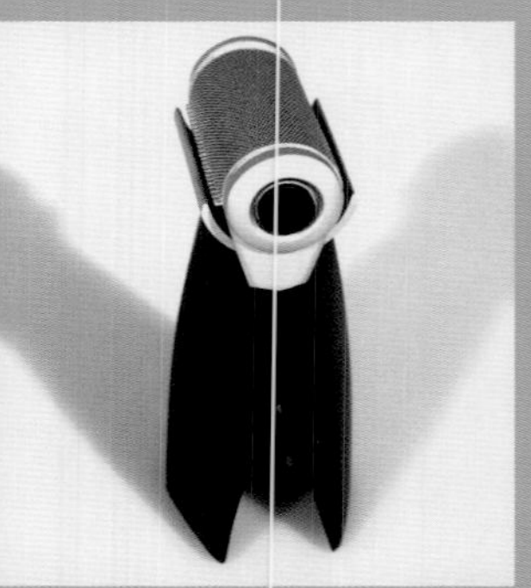

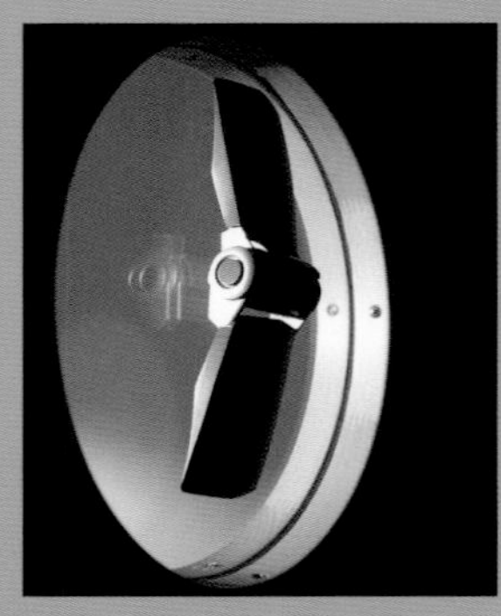

» in_suck_t «
» marek simko «

--- elektrisches rasiergerät -- energieversorgung über solarzellen -- um sonne zu tanken breitet "in_suck_t" die flügel aus und saugt sich an fensterflächen fest -- der scherkopf dreht sich mit hilfe des ratschenprinzips um die eigene achse und wird damit geschützt -- zielgruppe: - männer zwischen 21 und 35 jahren, im businessbereich tätig ---

--- electrical razor -- solar cells supply power -- to recharge, "in_suck_t" spreads its wings and is fastened to window surfaces with suction cups -- the head turns around its own axle on a ratchet principle, which protects it -- target group: - businessmen between the ages of 21 and 35 ---

» escape «
» frank rettenbacher «

--- "escape" versucht eine analogie zum altbekannten rasiermesser herzustellen -- durch den klappmechanismus wird der scherkopfteil geschützt, der shaver wird klein und handlich -- im scherkopfteil befindet sich ein linearmotor -- zielgruppe: - junge, designbewusste männer ---

--- "escape" attempts to establish an analogy to the old-fashioned straight razor -- a folding mechanism protects the head, the shaver is small, and there is a linear motor in the part containing the head -- target group: - young, design-conscious men ---

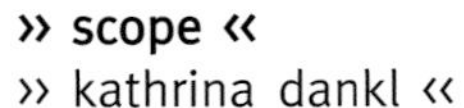

» scope «
» kathrina dankl «

--- "scope" ist ein internet-haaranalysesystem für den privatgebrauch -- vorgehensweise: - minikamera positionieren: 300-fache vergrösserung der haare, - kontakt zum hersteller wird über internet/e-mail hergestellt und - die bilddaten der haare werden übermittelt -- nach auswertung der haaranalyse empfiehlt die firma eine individuell zugeschnittene pflegeserie -- zielgruppe sind personen im alter von 20 bis 35 jahren, aus kreativberufen ---

--- "scope" is a hair analysis system for private use -- positioning of minicamera: 300 times enlarged shots of hair - contact with manufacturer by internet/e-mail and - transmission of visual data -- evaluation of analysis includes a tailor-made list of recommended hair care system – target group: people in creative professions between the ages of 20 and 35 ---

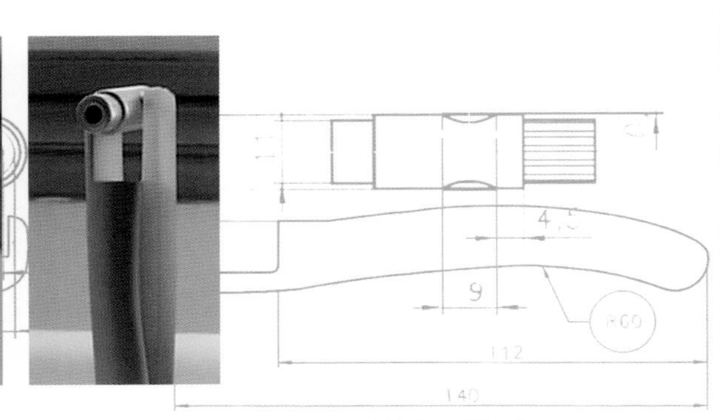

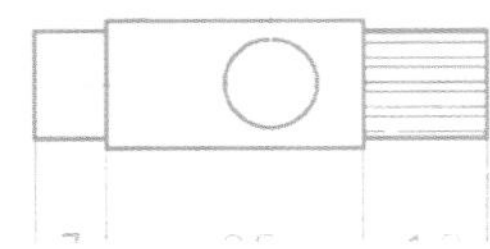

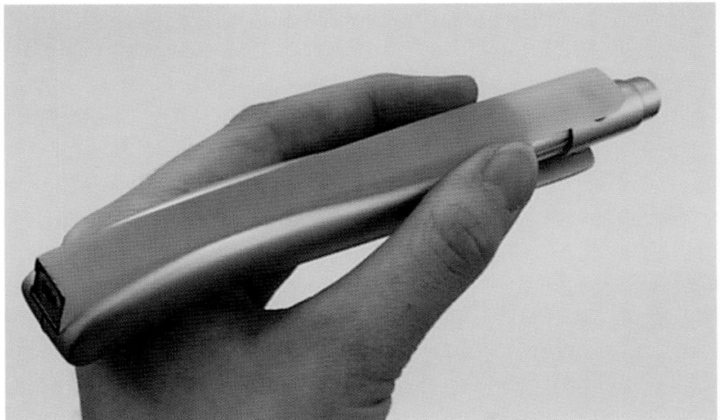

» waver «
» daniel zeisner «

--- der "waver" vereinigt zwei geräte in einer einheit, nämlich einen rasierer und einen ultraschallmassagekopf mit einem geltank -- man benötigt für die ultraschallmassage ein kontaktgel, das die schwingungen in die haut leitet -- das gel wird als aftershavelotion eingesetzt -- zielgruppe: - "business"-männer im alter zwischen 25 und 35 jahren ---

--- the "waver" combines two devices in one unit, a razor and an ultrasound massage head with a gel tank -- for the ultrasound massage, one needs ultrasound gel to transmit sound waves to the skin -- the gel is used as an after shave lotion -- target group: - businessmen between the ages of 25 and 35 ---

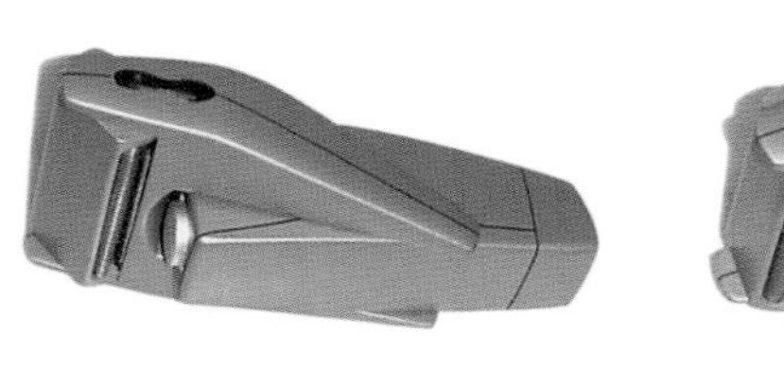

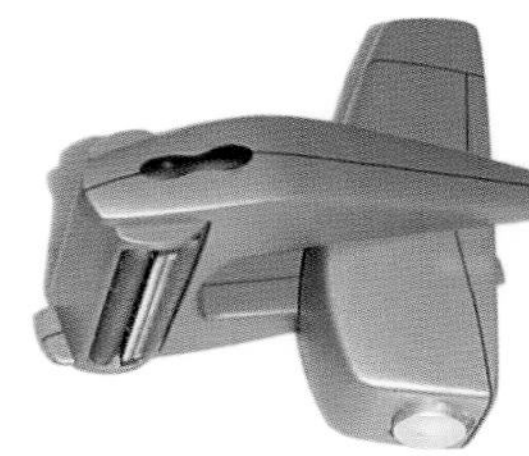

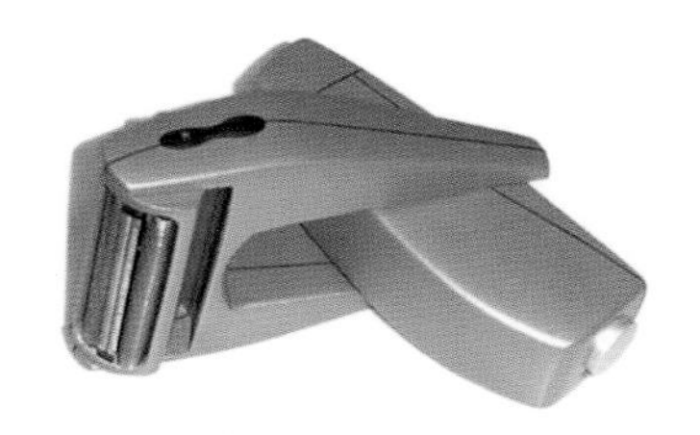

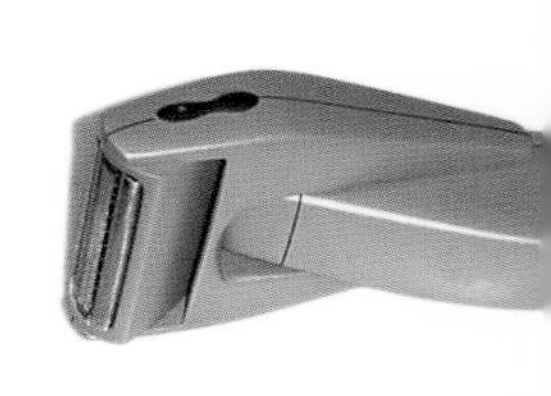

--- zielgruppe: - astronauten in weltraumstationen -- durch die integration einer absaugung in das haarschneidegerät, die mit einem staubsauger gekoppelt ist, wird das einatmen der haare unterbunden -- das gerät selbst kann man zwischen zeige- und mittelfinger klemmen ---

--- target group: - astronauts in space stations – the cut-off hair is vacuumed up with a small vacuum cleaner to prevent the hair that could be aspirated by astronauts, from floating around -- the device can be tucked between the index finger and the middle finger ---

» rasierer für den orbit «

» peter umgeher «

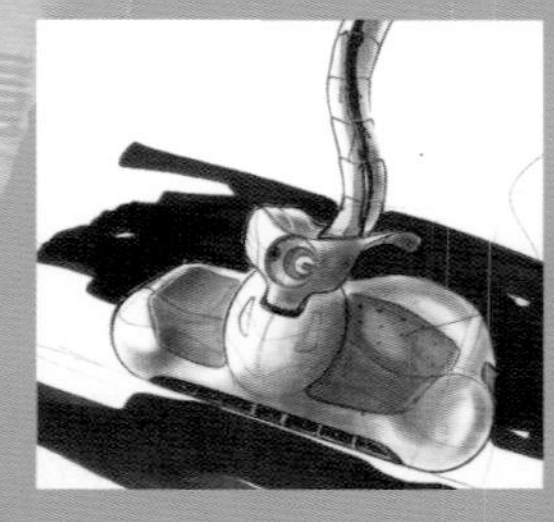

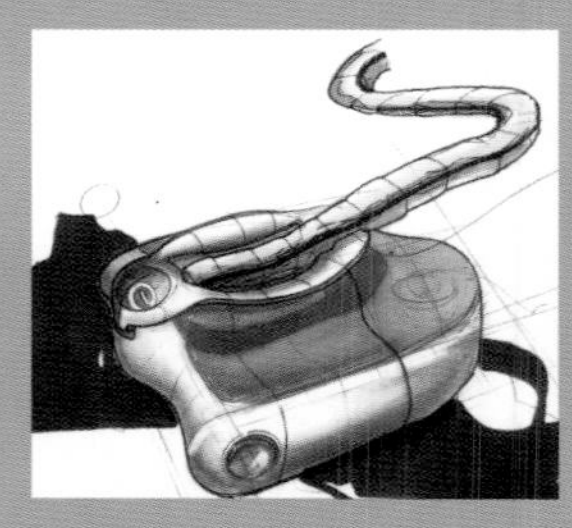

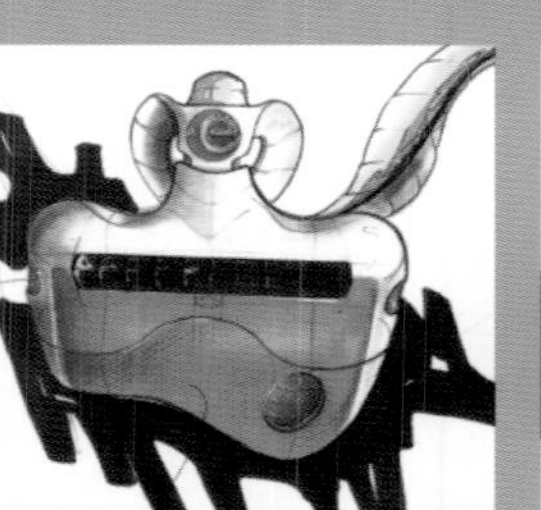

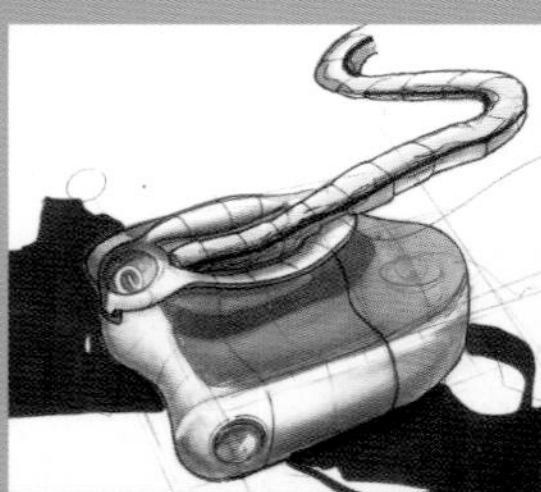

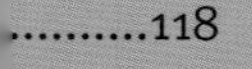

» spliner «

» lisa hampel «

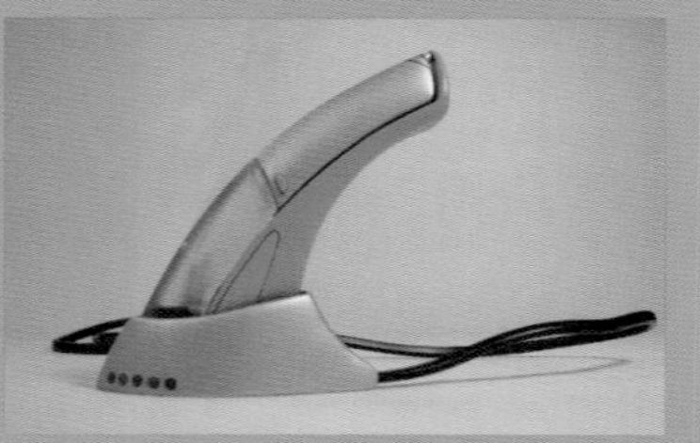

--- der "spliner" hat zwei funktionen: konturen vorzeichnen und konturen nachschneiden -- es wird wasserlösliche farbe auf das haar gesprüht -- mit einem kleinen rotor-schneide-element nachrasiert -- zielgruppe ist die "fun-generation" ---

--- "the spliner" has 2 functions: to mark out and shave outlines -- water-soluble coloring is sprayed on the hair -- the shaving is done with a small rotor cutting element -- target group is the "fun generation" ---

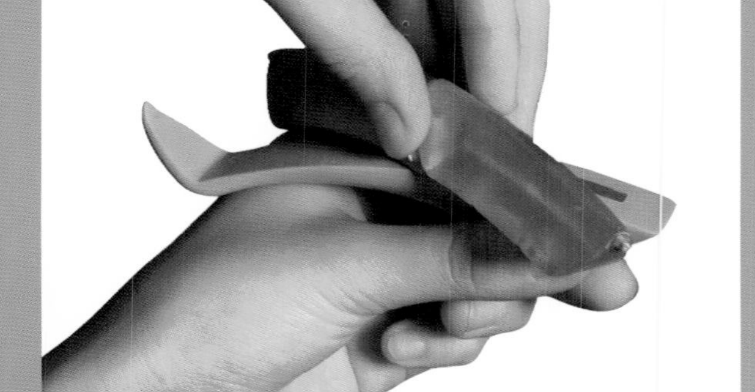

» cutcomb «
» gerald wirthenstätter «

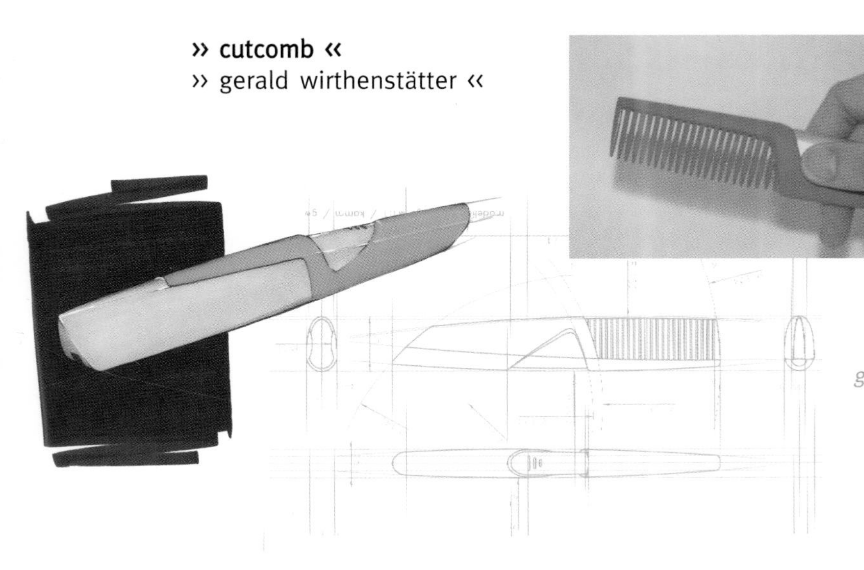

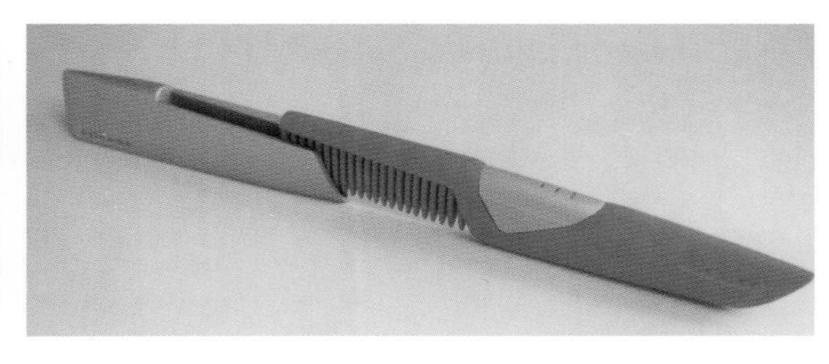

--- kamm und schere werden in einem gerät kombiniert -- die schneidblätter sind in den zähnen des kammes eingelassen und schneiden nach dem prinzip des langhaarschneiders bei einem rasierer -- zielgruppe: - "creative consumer" im alter von 25 bis 35 jahren mit sinn für lifestyle ---

--- comb and scissors are combined in one device -- the cutting blades are part of the teeth of the comb, and they cut using the same principle as the long hair trimmer in a razor -- target group: - "creative consumers" between the ages of 25 and 35 who have a sense for lifestyle ---

--- der "bird of prey" ist für die zielgruppe der extremsportler bestimmt -- er ist zusammenklappbar, robust und langlebig -- das geheimnis der offenen form ist der scherkopf -- in ihm befindet sich ein linearmotor, der die erwünschte horizontalbewegung der klingen ermöglicht ---

--- extreme sportsmen are the target group for the "bird of prey" -- it can be folded, it is robust and durable -- the secret of its open form is the head, which contains a linear motor that makes the desired horizontal movement of the blades possible ---

» bird of prey «
» elisabeth schwartz «

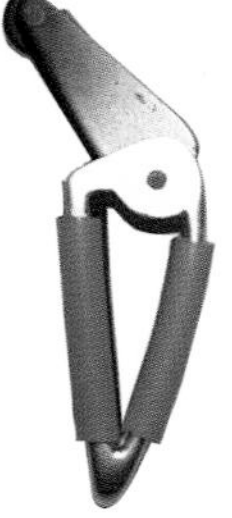

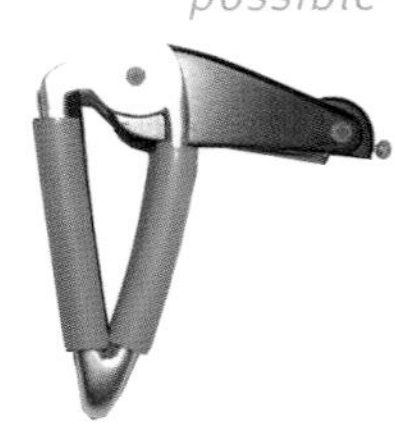

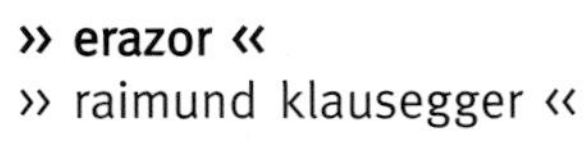

» erazor «
» raimund klausegger «

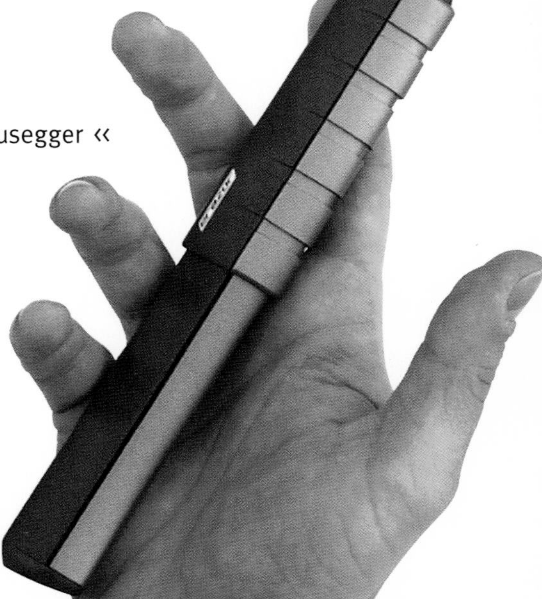

--- der keilförmige querschnitt des "erazors" ermöglicht ein genaues und schnelles rasieren -- der griffteil ist längs verschiebbar -- er schützt die scherfolie und dient auch als ein- ausschalter -- angetrieben wird er durch einen linearmotor, der von einem litium-ionen-akku mit energie versorgt wird -- zielgruppe: - männer mit vorliebe für trendsportarten ---

--- the wedge-shaped cross-section of the "erazor" makes precise and quick shaving possible -- the handle can be shifted lengthwise -- it protects the foil and also serves as off/on switch -- a lithium ion rechargeable battery provides the energy for the linear motor which, in turn, provides the power -- target group: - men who have a penchant for trendy sports ---

scherfolie *trimming foil*

gegenmesser *counterblade*

motor *motor*

langhaarschnitt *long hair trimming*

» i-cone «
» fritz pernkopf «

--- nasen- und ohrenhaartrimmer sowie rasierer in einem gerät vereint -- beide funktionen werden mit einem motor (rotierendes system) betrieben -- der nasen- und ohrenhaartrimmer wird im gerät versenkt -- zielgruppe: - männer zwischen 23 und 40 jahren, mit dem hang zu langlebigen produkten ---

--- nose and ear hair trimmer and razor in one single device -- both functions are operated by a motor (rotating system), and the ear hair trimmer is lowered into the device -- target group: - men between the ages of 23 and 40 who are partial to durable products ---

» blue line «
» johannes geisler «

--- ein trockenrasierer, der die energie des leitungswassers als antrieb verwendet -- doppelwandiger schlauch für zu- und abfluss des wassers -- im kopf des rasierers befinden sich düse und freistrahlturbine -- zielgruppe: - männliche jugendliche im alter von 13-19 jahren --- adolf loos staatspreis design 2001, anerkennungspreis

--- a dry razor, which uses the energy of tap water as a power source -- a double-walled hose for water inflow and drainage -- the razor head contains the jet and the impulse turbine -- target group: - young men between the ages of 13 and 19 --- adolf loos state prize for design 2001, honorable mention

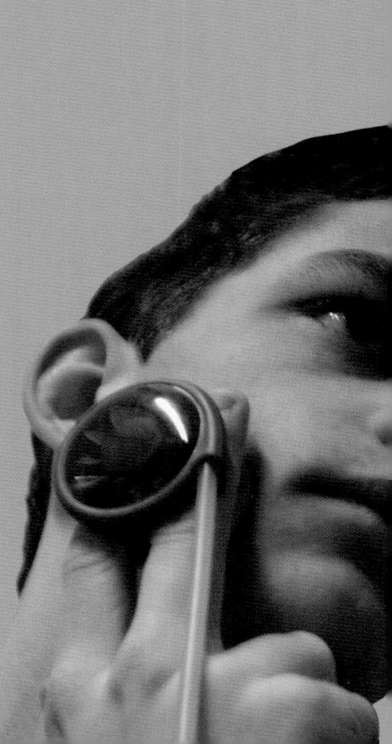

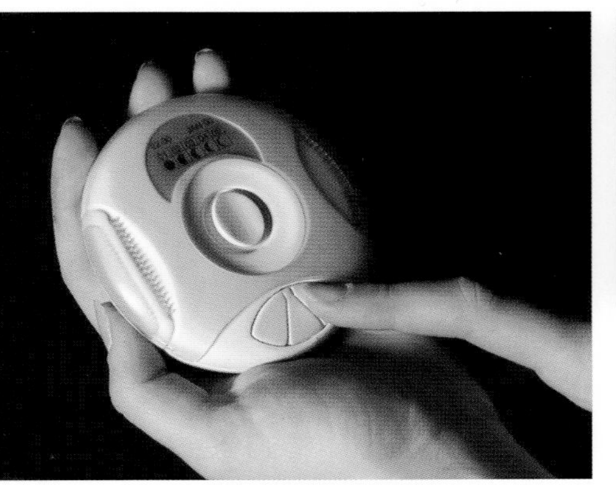

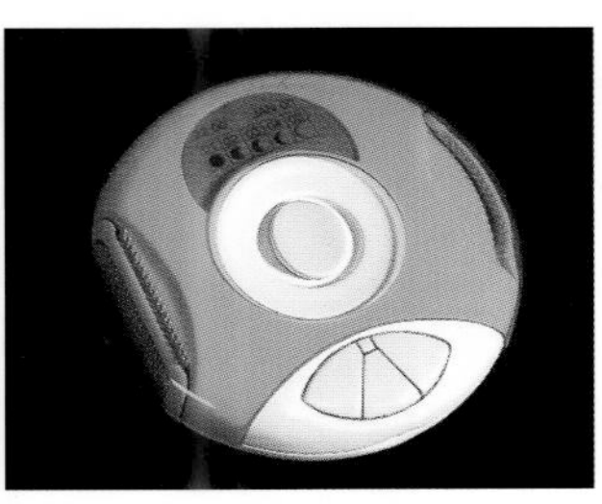

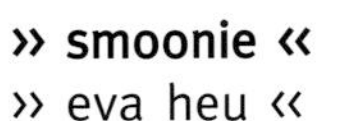

» smoonie «

» eva heu «

 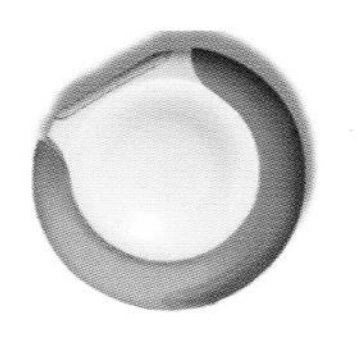 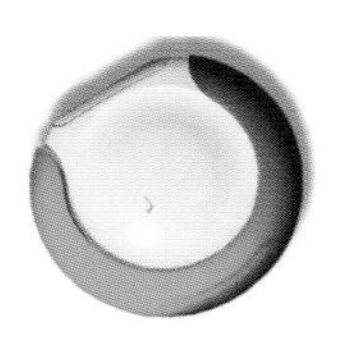

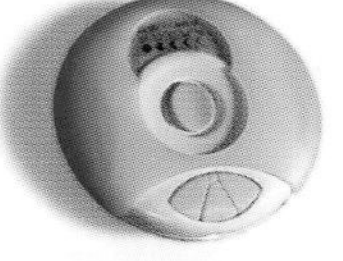  

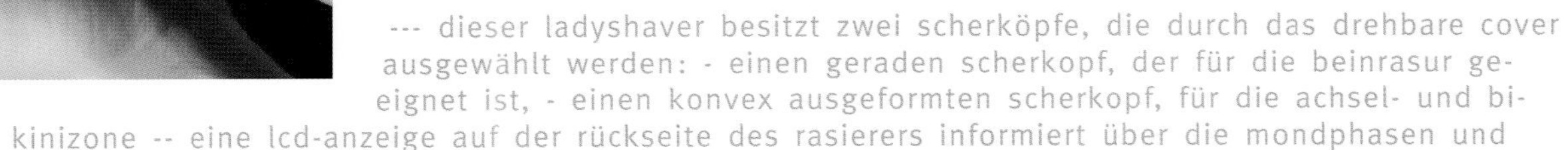

--- dieser ladyshaver besitzt zwei scherköpfe, die durch das drehbare cover ausgewählt werden: - einen geraden scherkopf, der für die beinrasur geeignet ist, - einen konvex ausgeformten scherkopf, für die achsel- und bikinizone -- eine lcd-anzeige auf der rückseite des rasierers informiert über die mondphasen und den biorhythmus -- auswechselbare cover in verschiedenen farben -- zielgruppen: - mädchen von 12-17 jahren, im produktumfeld von handy, mp3-player und disc-man ---

--- this lady's shaver has 2 heads, and you choose the head by turning the cover: - a straight head for shaving legs, - a head with a convex form for under the arms and the bikini area -- an lcd display on the back of the razor gives information about the moon phases and biorhythm -- interchangeable cover in various colors -- target groups: - girls between the ages of 12 and 17 who are also interested in products such as cell phone, mp3 player and discman ---

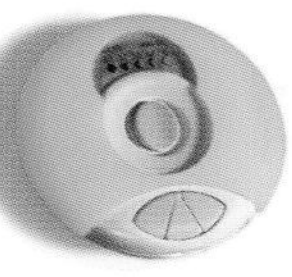  

 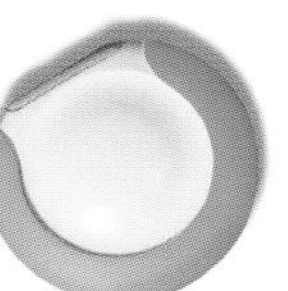 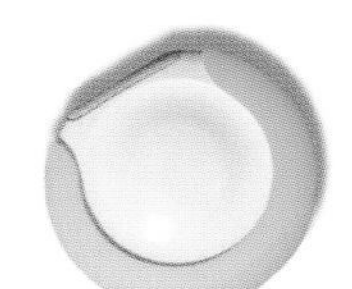

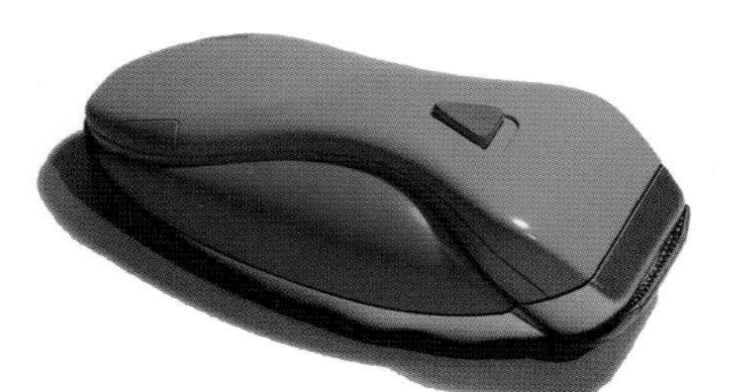

» bi-shaver «

» petra pucher «

--- grundidee: ein teilbarer ladyshaver -- für die rasur der beine stehen drei konkave scherflächen zur verfügung -- für punktgenaues rasieren in der achsel- und bikinizone ist der abnehmbare teil konvex geformt -- zielgruppen: - berufstätige oder - studierende frauen, im alter von 18 bis 30 jahren ---

--- the basic idea: a two-part lady's shaver -- three concave heads for shaving legs -- a removable convex part for precision shaving under the arms and for the bikini line -- target groups: - working women and - female students, between the ages of 18 and 30 ---

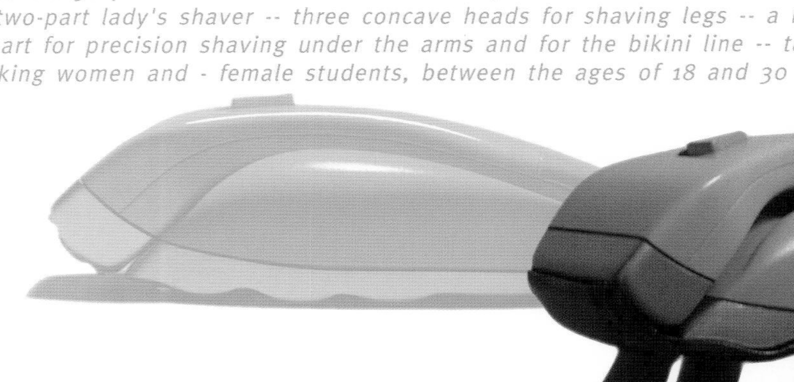

projekt: » trennen von holz und holzähnlichen werkstoffen «
7. semester/winter 2000/01

betreuer: hans peter a g l a s s i n g e r (teams design) - gastprofessor
georg w a g n e r - professor für engineering
werner k l e i s s n e r - walter l a c h – modellbau

thema: zielgruppen: heimwerker und profihandwerker. einstieg über analyse des ist-zustandes: erstens der heutigen arbeitsbedingungen und zweitens der heute im einsatz befindlichen geräte/werkzeuge (stichsäge, kreissäge, feinschnittsäge, fuchsschwanz, multisäge), daraus abgeleitet neue konzeptideen: verbesserung bzw. vereinfachung bestehender werkzeuge, geräte für spezielle anwendungen.

project: » separation of wood and materials similar to wood «
7th semester/fall 2000/01

advisors: hans peter a g l a s s i n g e r (teams design) - guest professor
georg w a g n e r - professor of engineering
werner k l e i s s n e r - walter l a c h - model building

subject: target groups: do-it-yourselfers and professional skilled workers. the first phase is an analysis of the actual situation. first, the current working conditions and secondly, the devices and tools, which are being used today (fret saw, circular saw, precision saw, handsaw, multisaw) and new concepts, which are derived from them, such as improvement and/or simplification of existing tools or tools and devices for special uses.

sign

projektarbeiten des fh-studienganges industrial design-graz

projects of the fh-degree course industrial design-graz

2000/01
(19)

holz

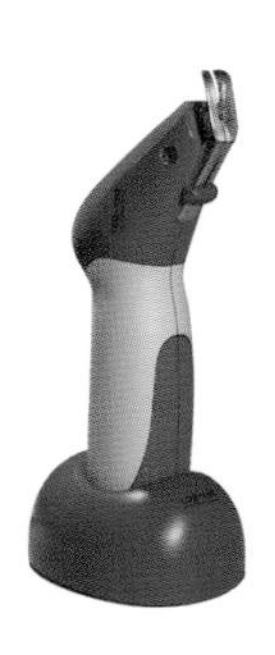

» short story - astsäge *branch saw* «

» albert ebenbichler «

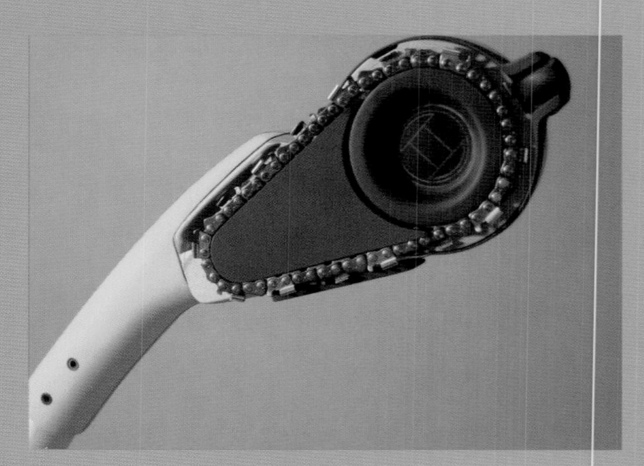

ruheposition = kettenschutz *resting position = chain protection*

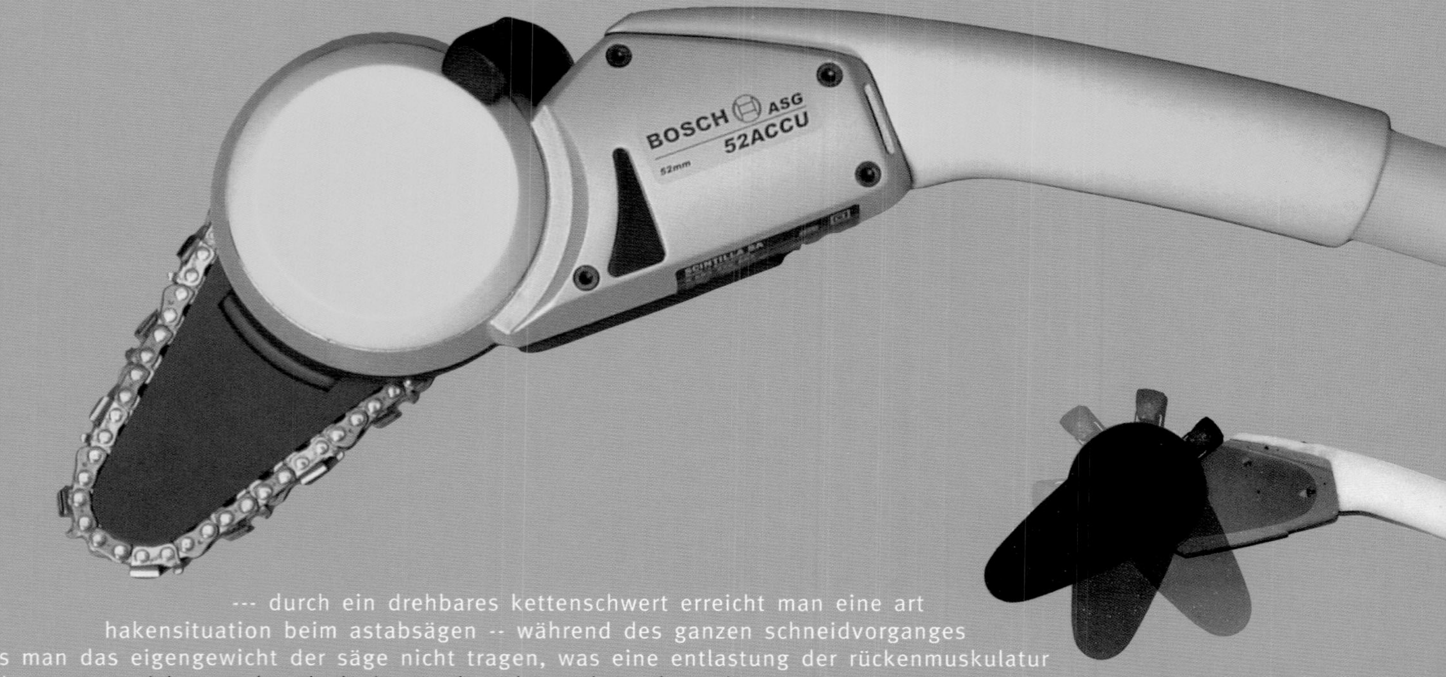

--- durch ein drehbares kettenschwert erreicht man eine art hakensituation beim astabsägen -- während des ganzen schneidvorganges muss man das eigengewicht der säge nicht tragen, was eine entlastung der rückenmuskulatur bewirkt -- ›auszeichnung: bosch design nachwuchswettbewerb workart 2001‹

--- a chain sword that can be swiveled produces a kind hook situation when sawing off branches -- during the entire sawing operation, one does not have to carry the saw's own weight and this makes it easier on the back --- ›award: bosch design novice competition workart 2001‹

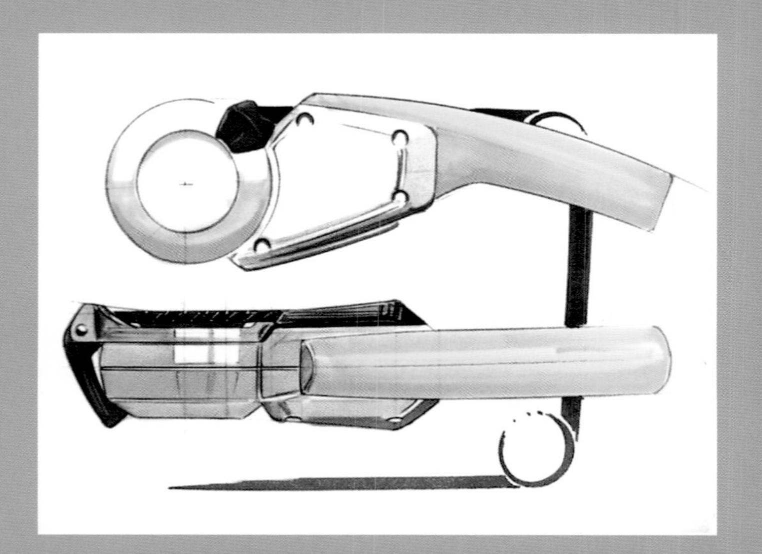

» h.a.p. lasercutter «

» daniel dockner «

--- der "h.a.p." ist ein lasercutter für tischler, modellbauer und restaurateure -- er schneidet furniere, holz- und kunststoffplatten und ist für intarsienarbeiten hervorragend geeignet -- das system besteht aus drei hauptelementen: - 1. baseunit: sie beinhaltet die laserquelle - 2. tableunit: sie ist von der baseunit abnehmbar und kann am arbeitstisch befestigt werden - 3. len (laser-pen): er besteht aus einer optischen einheit, inklusive sensoren, die den vorschub messen, um die laserleistung zu regulieren --- ›2. preis: bosch design nachwuchswettbewerb workart 2001‹

--- the "h.a.p." is a laser cutter for carpenters, model/mock-up builders and restorers -- it cuts veneers, wood and particle board, corian, formica, laminate materials and is very well suited for marquetry work -- the system consists of three main elements: - 1. base unit: it contains the laser source - 2. table unit: it can be removed from the base unit and can be fixed on a table - 3. len (laser pen): it consists of an optical unit, including sensors, which measure the thrust in order to regulate the laser power ---

›2nd prize: bosch design novice competition workart 2001‹

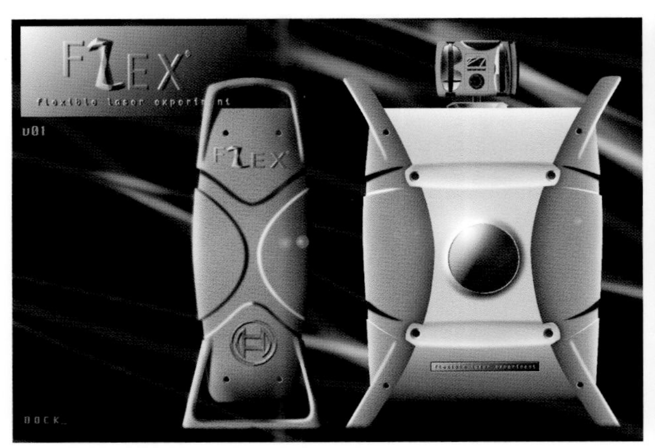

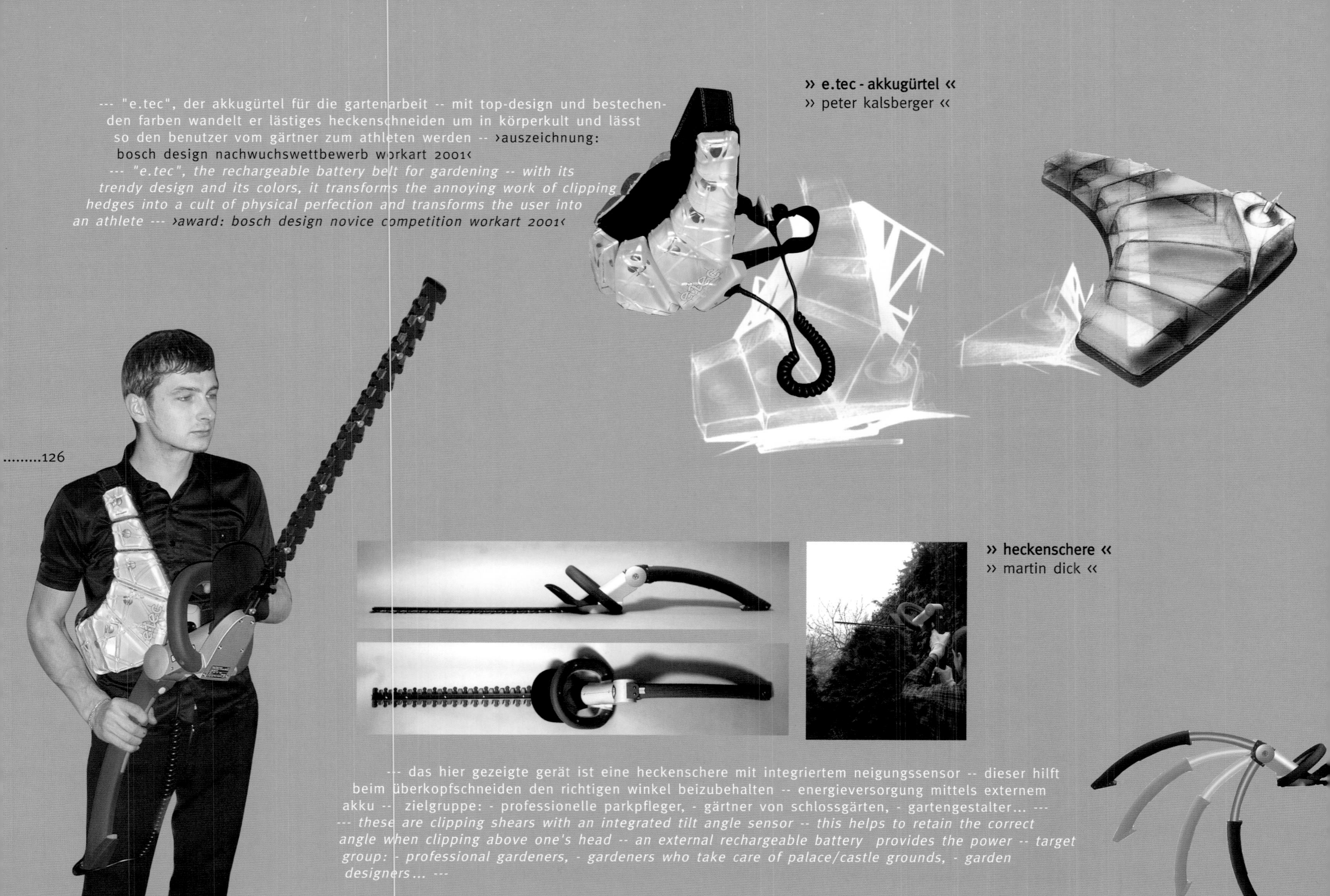

» e.tec - akkugürtel «
» peter kalsberger «

--- "e.tec", der akkugürtel für die gartenarbeit -- mit top-design und bestechenden farben wandelt er lästiges heckenschneiden um in körperkult und lässt so den benutzer vom gärtner zum athleten werden -- ›auszeichnung: bosch design nachwuchswettbewerb workart 2001‹

--- "e.tec", the rechargeable battery belt for gardening -- with its trendy design and its colors, it transforms the annoying work of clipping hedges into a cult of physical perfection and transforms the user into an athlete --- ›award: bosch design novice competition workart 2001‹

» heckenschere «
» martin dick «

--- das hier gezeigte gerät ist eine heckenschere mit integriertem neigungssensor -- dieser hilft beim überkopfschneiden den richtigen winkel beizubehalten -- energieversorgung mittels externem akku -- zielgruppe: - professionelle parkpfleger, - gärtner von schlossgärten, - gartengestalter... ---

--- these are clipping shears with an integrated tilt angle sensor -- this helps to retain the correct angle when clipping above one's head -- an external rechargeable battery provides the power -- target group: - professional gardeners, - gardeners who take care of palace/castle grounds, - garden designers... ---

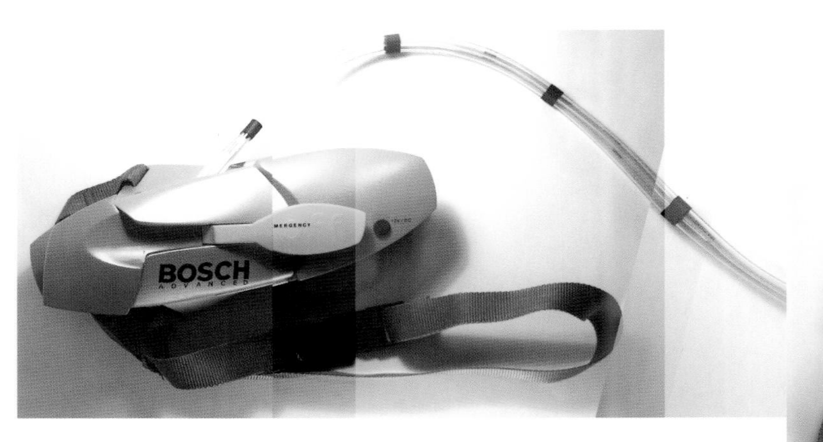

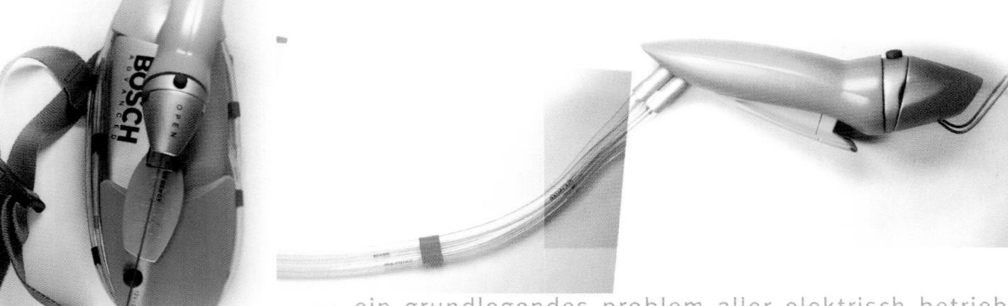

» hydraulik - gartenwerkzeug «

» mark ischepp «

--- ein grundlegendes problem aller elektrisch betriebenen gartenwerkzeuge ist ihr hohes gewicht, wodurch ein bequemes arbeiten auf dauer unmöglich wird -- daher trennung von gewicht (hydraulikpumpe) und funktion (hydraulikmotor) -- wechselsystem: ein auf hub- und zugbewegung basierendes hydraulisches antriebssystem für unterschiedliche werkzeugtypen ---

--- a basic problem of all electrically powered gardening tools is their heavy weight, which makes any kind of lengthy job uncomfortable -- this tool separates the weight (hydraulic pump) and the function (hydraulic motor) -- alternating system: a hydraulic propulsion system, which is based on a lifting and pulling movement, for various kinds of tools ---

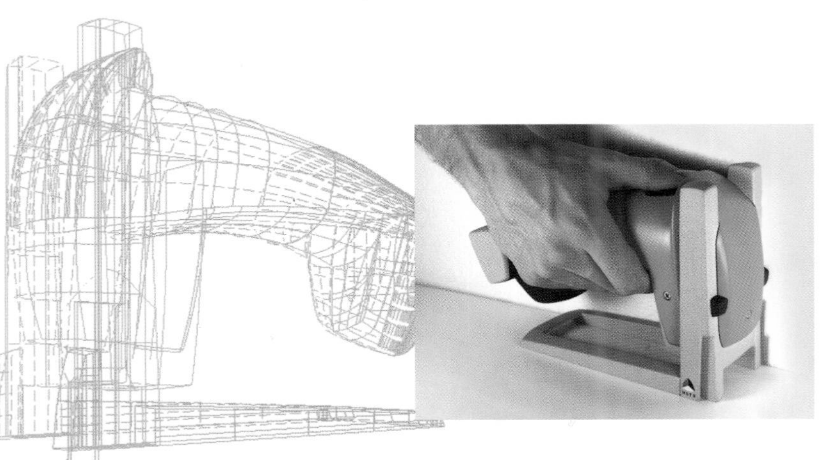

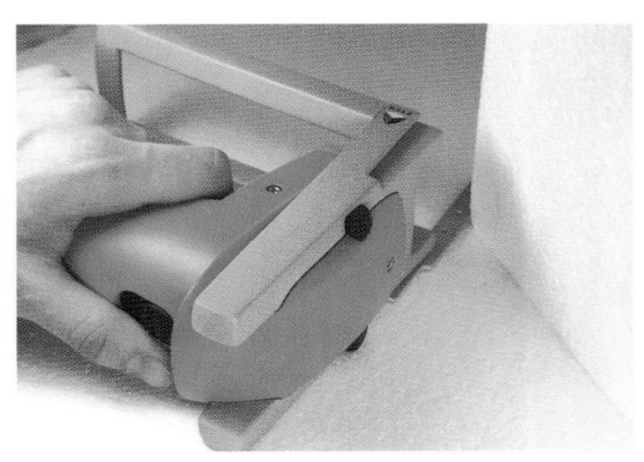

» etech «

» karin krumphals «

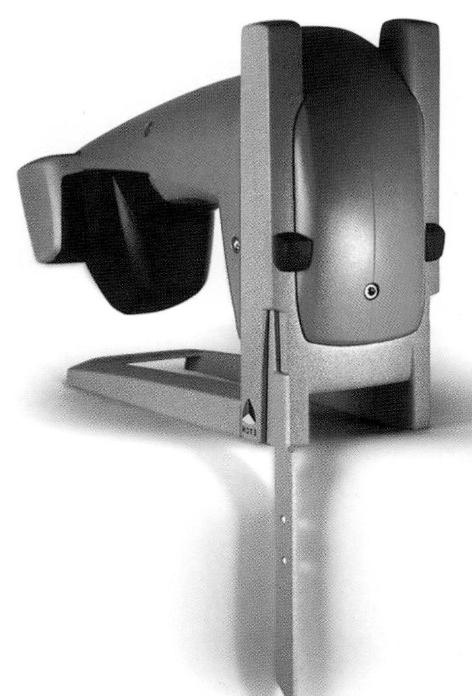

--- das konzept für diese säge war, die funktionen von pendelhubstichsäge und feinschnittsäge in einem werkzeug zu vereinen -- weiter ist diese säge in der lage, flächenbündige schnitte auszuführen sowie schnitte in die ecke -- zielgruppe: - der professionelle arbeitsbereich, wie beispielsweise ein tischler auf möbelmontage ---

--- the concept for this saw was to combine the functions of a self-aligning lifting compass saw and a fine saw in one tool -- further, this saw is able to cut flush with the surface and in corners -- target group: - professionals, for instance, carpenters installing built-in furniture ---

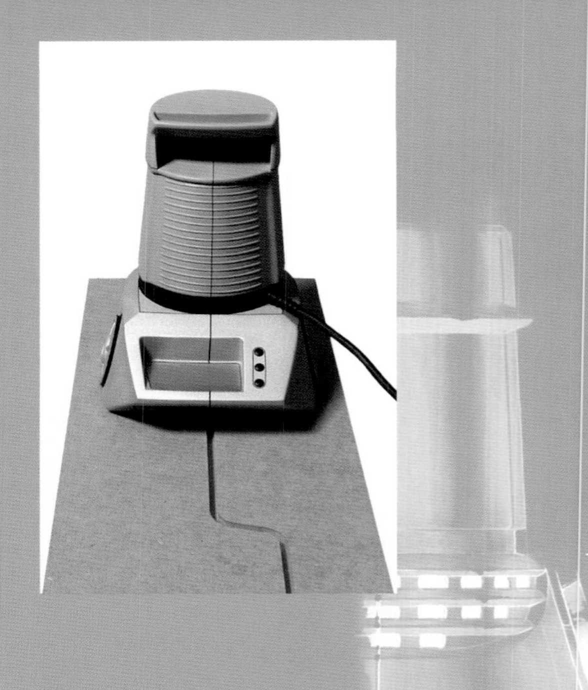

» shaper - cnc-fräser «
» joe lintl «

-- der "shaper" ist ein leistungsfähiger fräsroboter zum trennen von holzplatten aller art -- die zu bearbeitenden platten werden auf dem boden auf latten gelegt, und dann vom "shaper" mittels cad-programmierung getrennt, oder das schnittmuster wird mittels eines markierstiftes direkt auf der platte aufgetragen und der fräsroboter verfolgt diese linie mittels fotozelle -- zielgruppe: - kulissenbauer, - messestandbau, - modellbau ... --- ›3. preis: bosch design nachwuchswettbewerb workart 2001‹

--- the "shaper" is a powerful milling robot for cutting wooden boards -- the boards to be worked on are placed on the floor on slats and then cut by the "shaper" using cad programming, or the cutting pattern is traced directly onto the board with the marker and the milling robot follows this line using a photo cell -- target group: - makers of theater sets, - exhibition booths, - models and mock-ups ... --- ›3nd prize: bosch design novice competition workart 2001‹

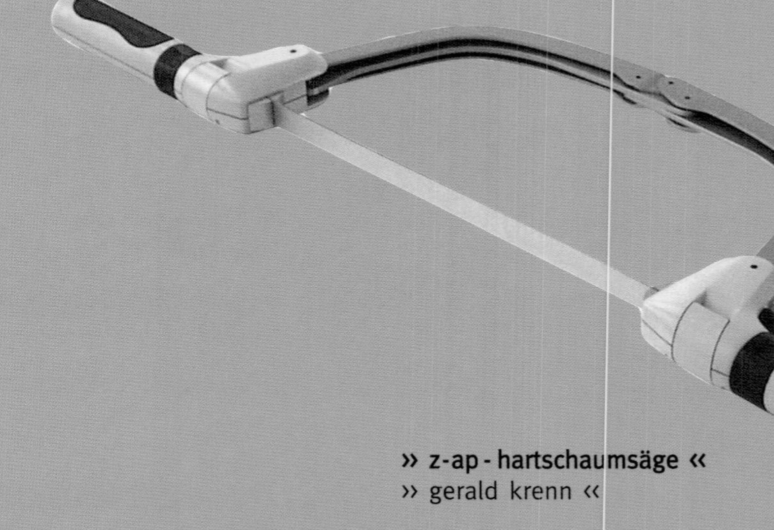

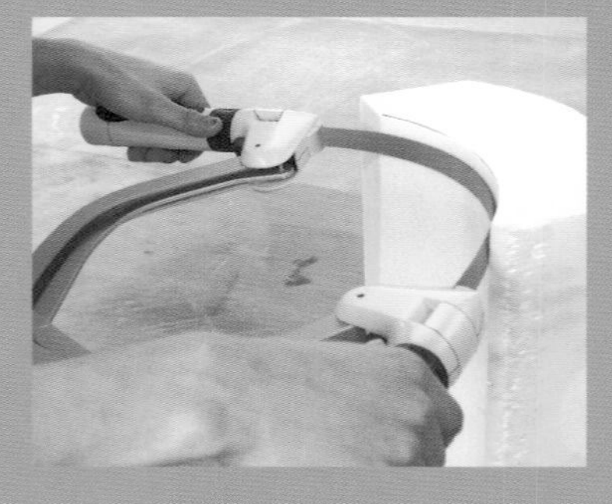

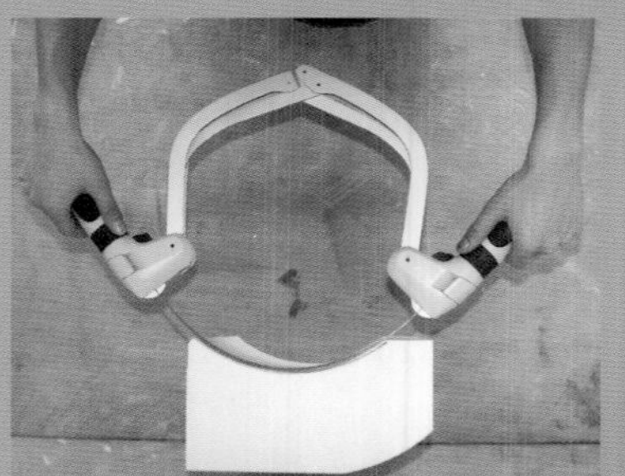

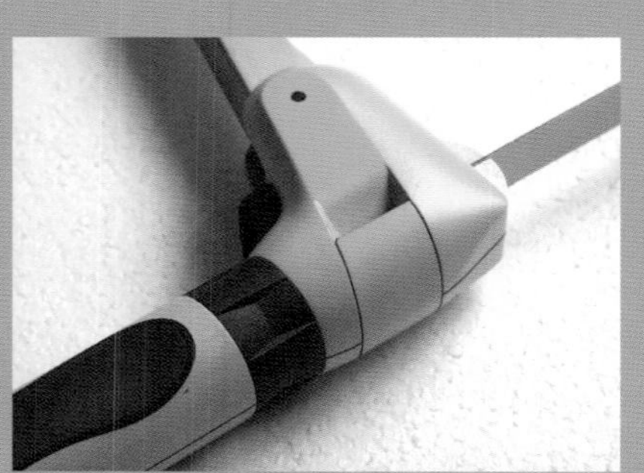

» z-ap - hartschaumsäge «
» gerald krenn «

--- diese spezielle hartschaumsäge ist für modellbauer und künstler konzipiert -- sie ermöglicht es, kurven und bögen zu schneiden, und so schnell positiv oder negativ gespannte flächen eines werkstücks zu bearbeiten -- betrieben wird die säge von einem 12-volt-motor, der samt getriebeeinheit im rechten hangriff sitzt ---

--- this special hard foam saw is designed especially for model and mock-up makers and artists -- it makes it possible to cut curves and arches and to quickly work on positively or negatively stretched surfaces of a workpiece -- the saw is powered by a 12 volt motor, which is contained in the right handle together with the gearshift ---

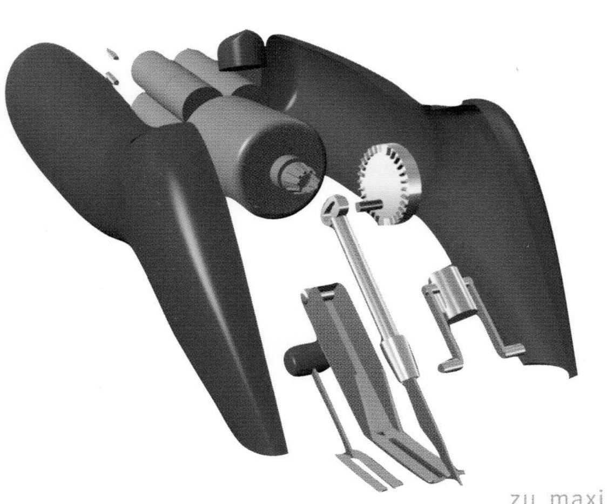

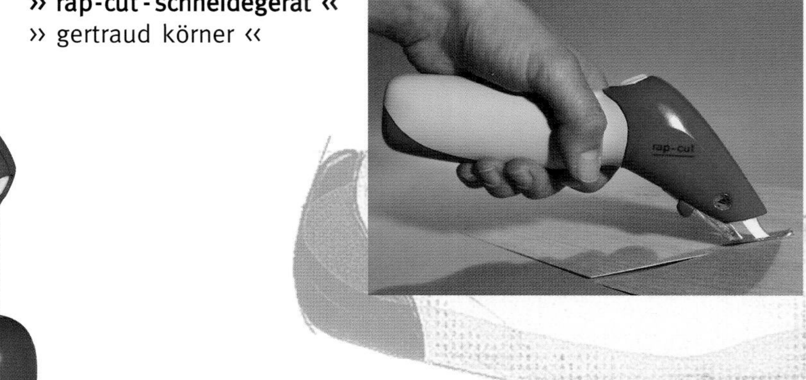

» rap-cut - schneidegerät «
» gertraud körner «

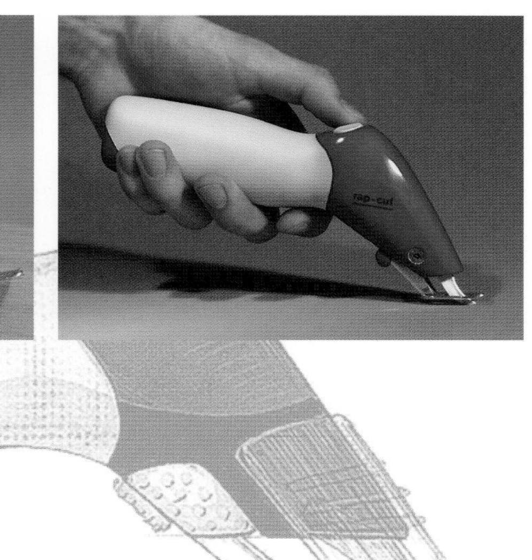

--- zum schneiden von plattenmaterialien bis zu maximal 5 mm (karton, furnier, schaumstoff, usw.) werden schneidemesser verwendet, die einen gewissen kraftaufwand erfordern -- ziel ist es, ein elektrisches schneidegerät zu entwickeln, welches eine austauschbare klinge hat, die sich gegen ein füsschen oszilierend auf und ab bewegt ---

--- utility knives, which need a certain amount of strength, are used to cut boards up to a maximum thickness of 5 mm (cartons, veneer, plastic foam, etc.) -- the goal is to develop an electrical cutting tool, which has an interchangeable oscillating blade ---

» zappa - holzspalter «
» beate dörflinger «

--- "zappa" trennt das holzscheit sicher, sauber und leise in zwei teile -- "zappa" kann nur mit geschlossenen türen in betrieb gesetzt werden -- das messer ist beidseitig an den hydraulikzylindern befestigt und wird so durch das holz gedrückt -- "zappa" ist eine neue art von werkzeug, das sich voll in den wohnbereich integriert ---

--- "zappa" splits the wood safely, cleanly, and quietly -- "zappa" can be used only if the doors are closed -- the blade is fixed on both sides to the hydraulic cylinders and is pushed through the wood -- "zappa" is the kind of tool that can be fully integrated into the interior living area ---

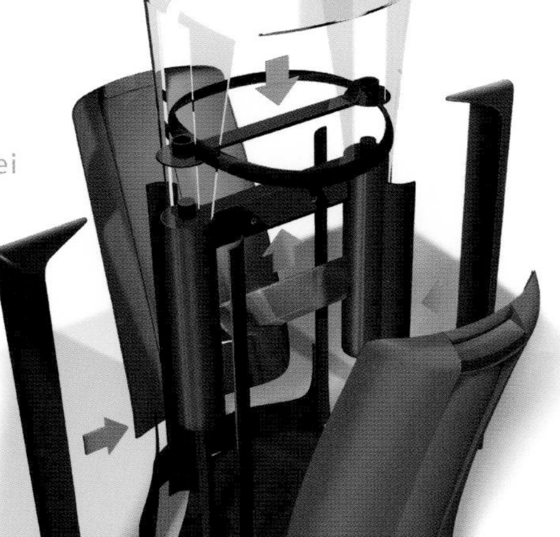

FH JOANNEUM
FACHHOCHSCHULSTUDIENGÄNGE
DER TECHNIKUM JOANNEUM GMBH

2000/01 (20)

projekt: » wasser: fun + motion «
4. semester/sommer 2001

betreuer: kurt h i l g a r t h (fancy form) - gastprofessor
georg w a g n e r - professor für engineering
werner k l e i s s n e r - walter l a c h - modellbau

thema: der rahmen für diese aufgabenstellung ist wiederum bewusst sehr weit gesteckt: nachdem vor einem jahr der wintersport mit "bewegung auf schnee und eis" gefragt war, dreht es sich diesmal um das sommervergnügen an/in seen oder am/im meer. es geht also um die entwicklung von innovativen wasser(sport)geräten. zielgruppen: alle, die es in ihrer freizeit oder im urlaub ans wasser zieht – von spielenden kindern bis zu passionierten wassersportlern.

project: » water: fun + motion «
4th semester/spring 2001

advisors: kurt h i l g a r t h (fancy form) - guest professor
georg w a g n e r - professor of engineering
werner k l e i s s n e r - walter l a c h - model building

subject: the planned scope for this task was deliberately very broadly-based. a year ago, winter sports with the "locomotion on snow and ice" project was in demand; this time, the subject is summer fun at/in lakes or at/in the ocean. it is all about the development of innovative water (sports) devices. target groups are anyone who is attracted to water in their leisure time or for their vacation – from playing children to passionate water sports enthusiasts.

sign

projektarbeiten des fh-studienganges industrial design-graz

projects of the fh-degree course industrial design-graz

2000/01
(20)

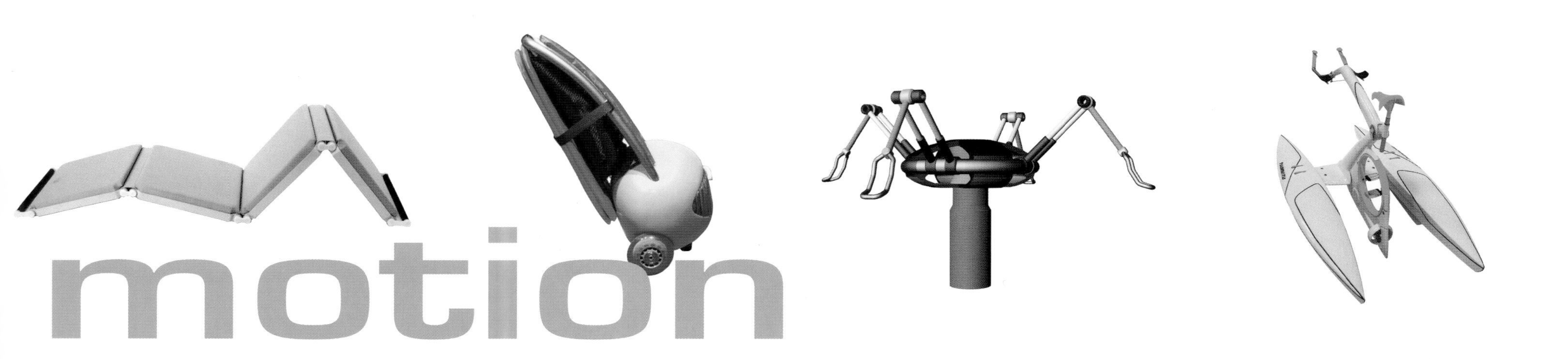

» tuamotu «

» raimund klausegger «

--- "tuamotu" ist ein ein-personen-renntretboot mit besonders kompakten abmessungen -- beide bootsrümpfe lassen sich abnehmen -- alle verschlüsse sind leicht zu bedienende schnellspanner ---

--- "tuamotu" is a one-man racing pedal boat with particularly compact measurements -- both boat hulls can be removed -- all fasteners are easy to use quick-action turnbuckles ---

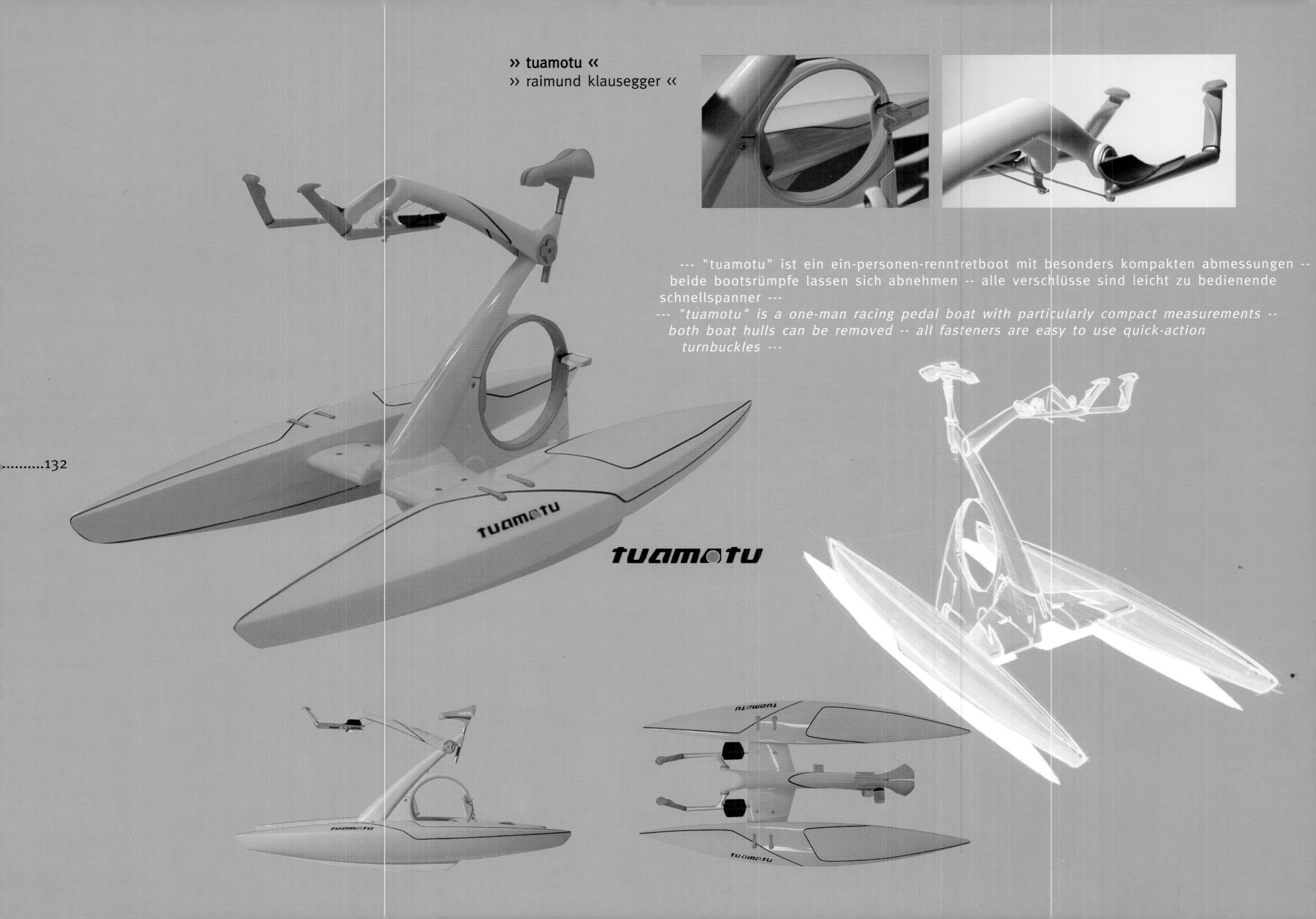

» bellyhoo «
» kathrina dankl «

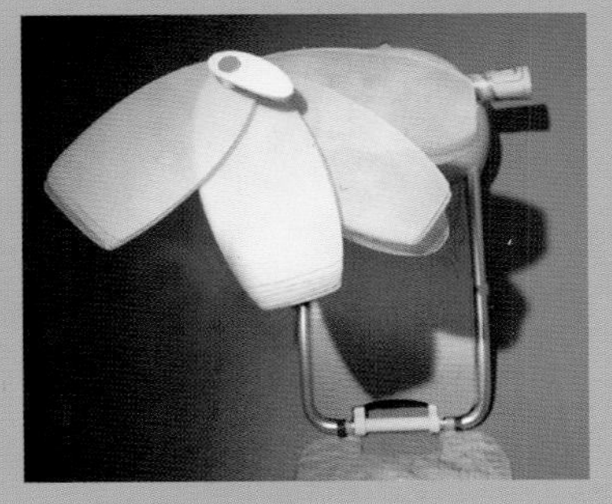

--- "ballyhoo" ist ein trolley mit integrierter liege und sonnenschutz -- ein komplettunit für die geniesser am strand, die entspannung suchen bei lesen, musik hören, sonnen, schlafen, telefonieren -- auf diese zielgruppe zugeschnitten ist der aufbau von "bellyhoo" mit seiner klaren trennung zwischen hardbox für bücher, walkman, zeitschriften und getrennt davon einer luftdurchlässigen softbox für handtücher, badeanzug etc. -- so bleiben die trockenen sachen geschützt ---

--- "bellyhoo" is a trolley with an adjustable mat and sunshade -- a completunit for pleasure seekers at the beach by reading, listening to music, basking, sleeping, phoning -- there is a clear separation between a hard-sided box for books, portable cd-player and cell phone and a soft-sided container that is pervious to air for towel and bathing suit -- so "bellyhoo" guarantees that everything stays dry ---

--- wasser ist das ideale medium für eine landefläche ohne verletzungsgefahr -- wirf dich ins element wasser -- das projekt ist inspiriert vom karussell, mit dem grossen vorteil, das gerät sorgenlos verlassen zu können -- die kragarme sind austauschbar, sodass eine fixe und eine flexible aufhängungsvariante möglich ist -- konzipiert für kinder, die nach technoidem spass suchen ---

--- water as an ideal medium for landing without getting hurt -- throw yourself into the element water -- concept "cyclone" is a roundabout conceived for an amusement park and can be used by four brave children at the same time -- the arm consoles are interchangeable to provide either a fixed or a flexible handle position -- designed for children, who are searching for technoid fun ---

» cyclone «
» marek simko «

--- bei "unit" handelt es sich um schwimmende sechseckige elemente, welche durch speziell ausgeformte clipelemente untereinander verbunden werden können -- die dose mit schraubverschluss in der mitte dient als wasserdichte aufbewahrungsmöglichkeit ---

--- "unit" is a floating hexagonal element, the parts of which can be connected among each other by specially shaped clips -- the container with screw-on top, which is located in the center, can be used as waterproof storage ---

» unit «
» peter umgeher «

--- "visioneer" soll funktionelle und formale innovation in den markt der taucherbrillen bringen -- eine voll gekrümmte sichtscheibe ermöglicht einen verzerrungsfreien panoramablick (neue technologie von hydrooptik) -- zusätzlich hat "visioneer" eine integrierte digicam für bilder und kurzfilme ---

--- "the visioneer" differs from regular dive masks both in form and in function -- as opposed to regular dive masks, the "visioneer" has curved lenses (new technology by hydrooptix), which make a distortion-free panoramic view possible -- besides, there is a digicam integrated into the "visioneer", which can make both photographs and short videos ---

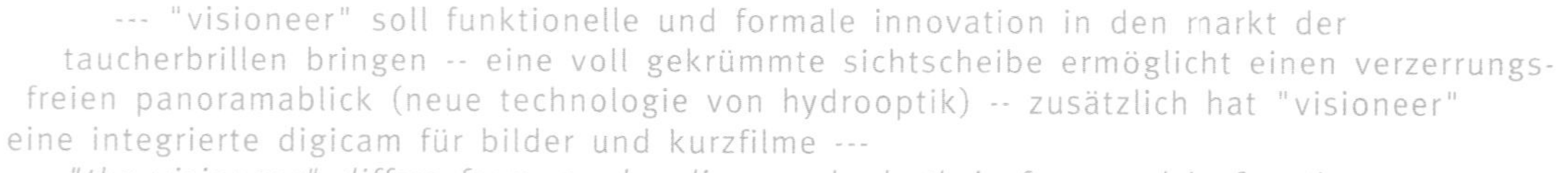

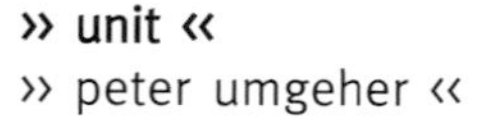

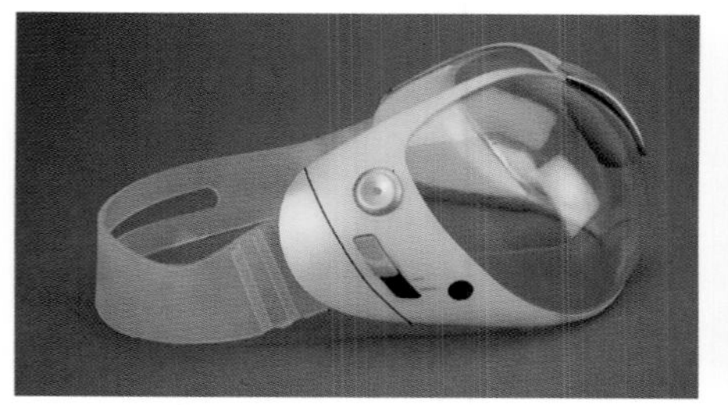

» visioneer «
» lisa hampel «

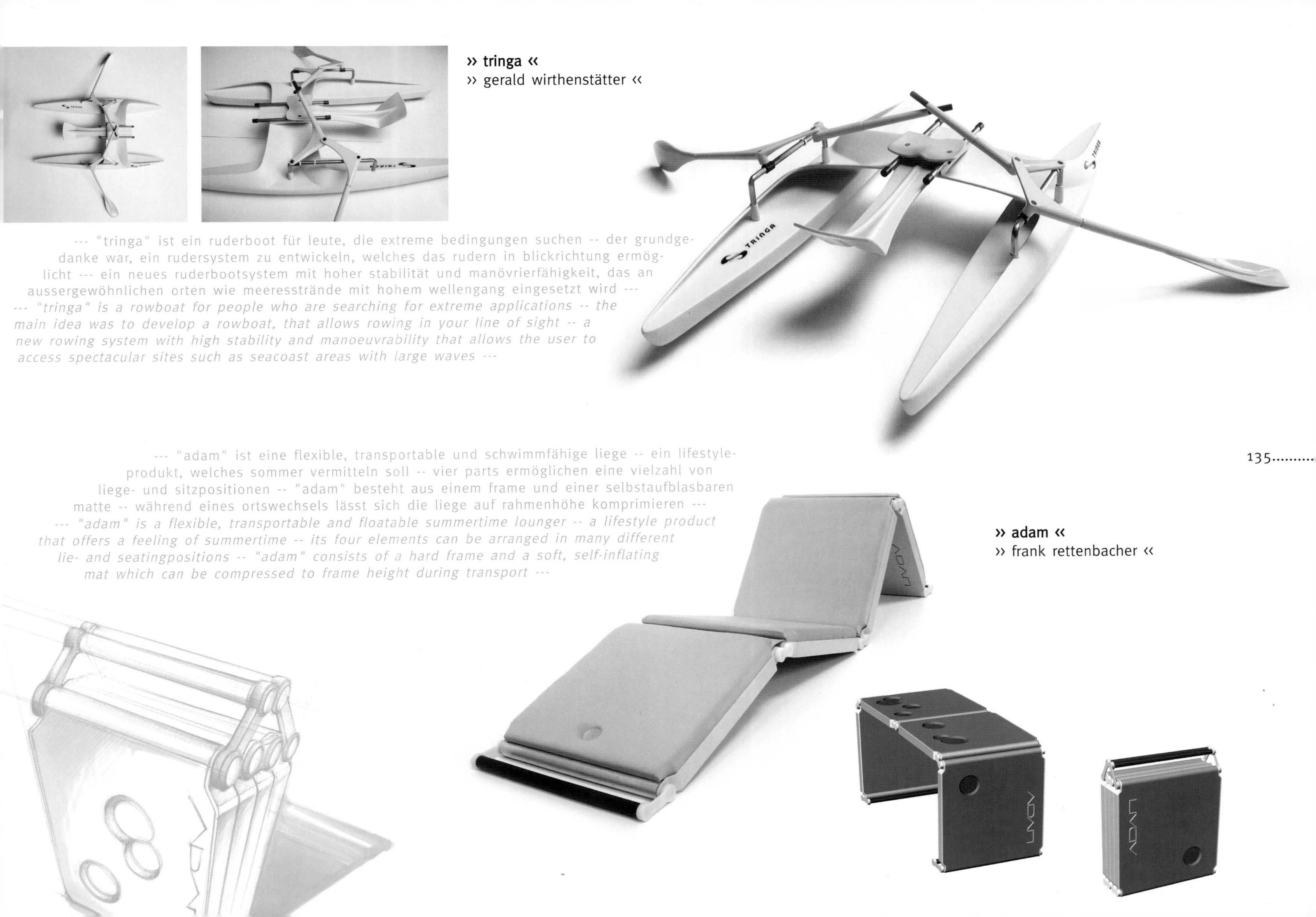

›› tringa ‹‹
›› gerald wirthenstätter ‹‹

--- "tringa" ist ein ruderboot für leute, die extreme bedingungen suchen -- der grundgedanke war, ein rudersystem zu entwickeln, welches das rudern in blickrichtung ermöglicht --- ein neues ruderbootsystem mit hoher stabilität und manövrierfähigkeit, das an aussergewöhnlichen orten wie meeresstrände mit hohem wellengang eingesetzt wird ---

--- "tringa" is a rowboat for people who are searching for extreme applications -- the main idea was to develop a rowboat, that allows rowing in your line of sight -- a new rowing system with high stability and manoeuvrability that allows the user to access spectacular sites such as seacoast areas with large waves ---

--- "adam" ist eine flexible, transportable und schwimmfähige liege -- ein lifestyle-produkt, welches sommer vermitteln soll -- vier parts ermöglichen eine vielzahl von liege- und sitzpositionen -- "adam" besteht aus einem frame und einer selbstaufblasbaren matte -- während eines ortswechsels lässt sich die liege auf rahmenhöhe komprimieren ---

--- "adam" is a flexible, transportable and floatable summertime lounger -- a lifestyle product that offers a feeling of summertime -- its four elements can be arranged in many different lie- and seatingpositions -- "adam" consists of a hard frame and a soft, self-inflating mat which can be compressed to frame height during transport ---

›› adam ‹‹
›› frank rettenbacher ‹‹

projekt: » fahrzeuge für studenten «
6. semester/sommer 2001

betreuer: gerhard h e u f l e r - gerald k i s k a - professoren für design
georg w a g n e r - professor für engineering -
werner k l e i s s n e r - walter l a c h - modellbau -

thema: in den 60er jahren war die frage nach dem bevorzugten studentenfahrzeug noch klar zu beantworten: favorit war der citroen 2 cv. heute stellt sich die situation wesentlich komplexer dar: vom skateboard über das mountainbike bis zum gebrauchten golf cabrio, vom steyr-waffenrad bis zur bevorzugung öffentlicher verkehrsmittel reicht die oft widersprüchliche palette. deshalb die frage: wie sehen studenten die studentenmobilität von morgen?

project: » vehicles for students «
6th semester/spring 2001

advisors: gerhard h e u f l e r - gerald k i s k a - professors of design -
georg w a g n e r - professor of engineering -
werner k l e i s s n e r - walter l a c h - model building -

subject: in the 1960's the question of the prefered student vehicle was easy to answer – the favorite was the citroen 2cv. today, the situation is much more complex. the range of preferences that are often full of contradictions goes from the skateboard to the mountain bike to a used golf convertible, from a steyr waffenrad bicycle to a preference for public transportation. the question here is how do students today perceive the student mobility of tomorrow?

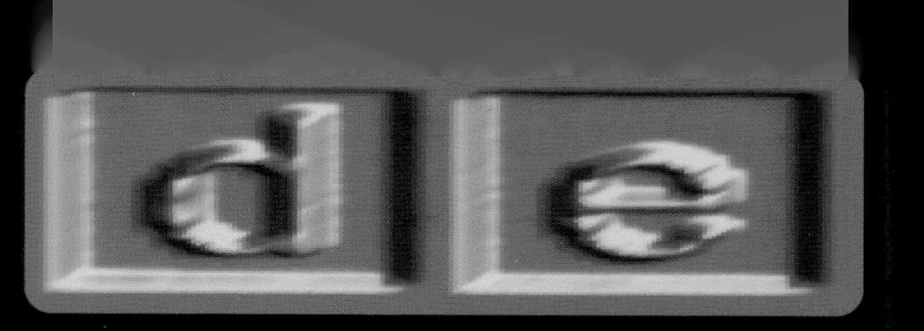

sign

projektarbeiten des fh-studienganges industrial design-graz
projects of the fh-degree course industrial design-graz

2001
(21)

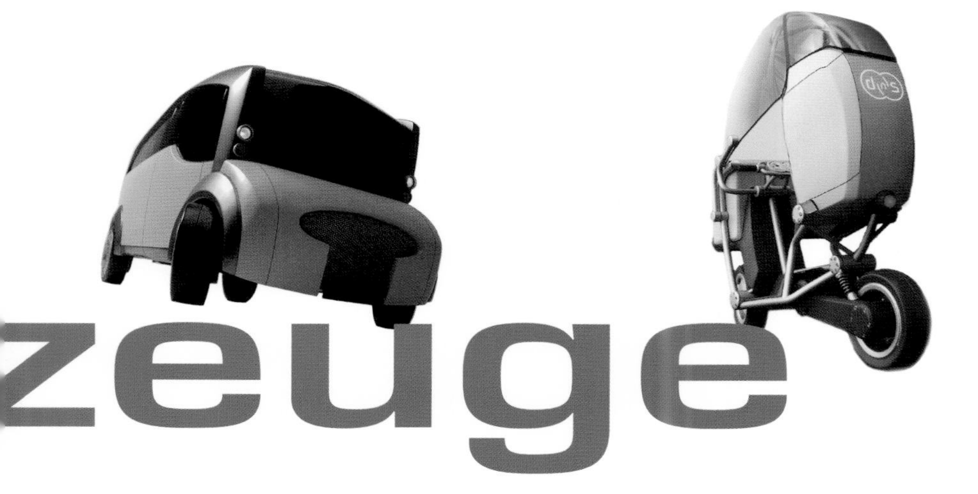

zeuge

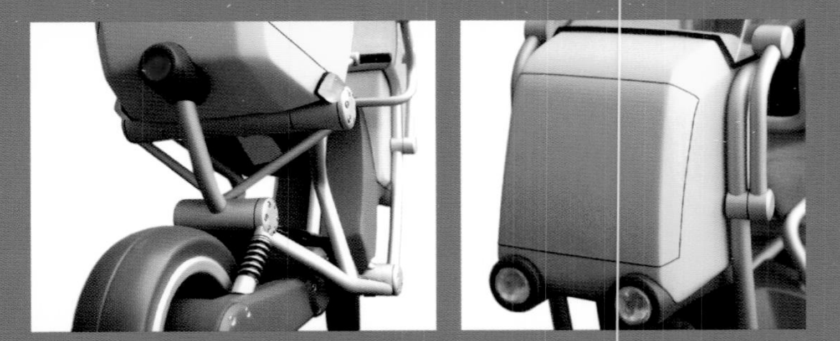

» e.c(2) - electric-convertible-2-wheeler «

» alexandra giselbrecht, robert hitthaler, andrea stelzer «

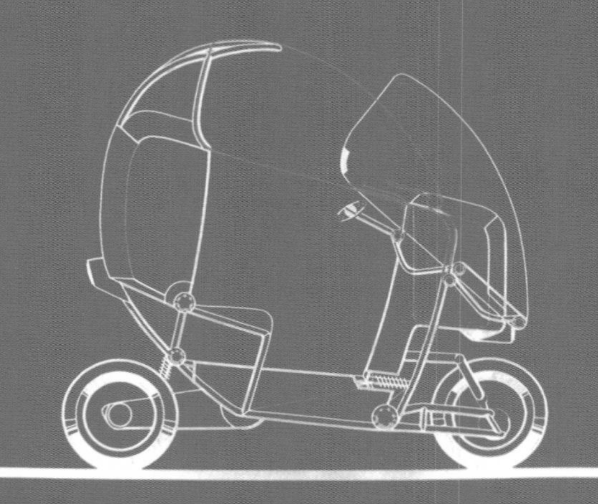

--- teil eines sharing fuhrparks, der auf infrastrukturelle bedürfnisse von studenten/innen abgestimmt ist -- kann mit einer chipkarte im ganzen stadtgebiet in betrieb genommen werden -- ohne helm fahrbar -- wandelbares erscheinungsbild (flexibler wetterschutz) -- schwerpunkt auf transportvolumina (getränkekiste im vorderen bereich, rucksäcke, einkäufe,... im rückenbereich) -- max. geschwindigkeit von 45km/h und reichweite von 30 km ---

--- the vehicle is part of a sharing fleet for students infrastructural demands -- will be available with a key-card in the entire city area -- can be ridden without helmet -- provides flexible protection against the elements -- focus on loading capacity (case of bottles in the front, backpacks, shopping bags,... in the rear) -- max. speed: 45 km/h -- max range: 30 km ---

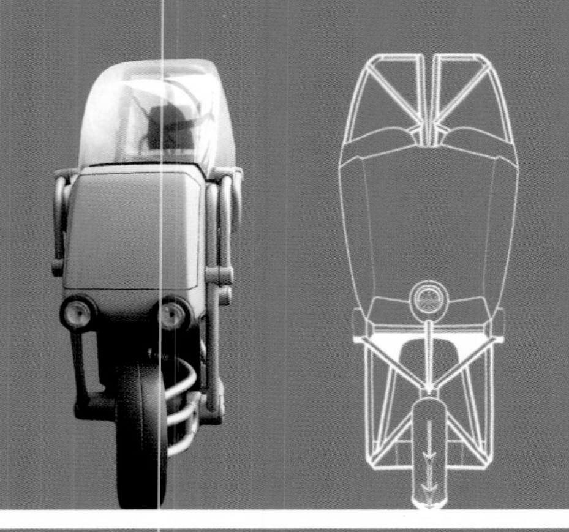

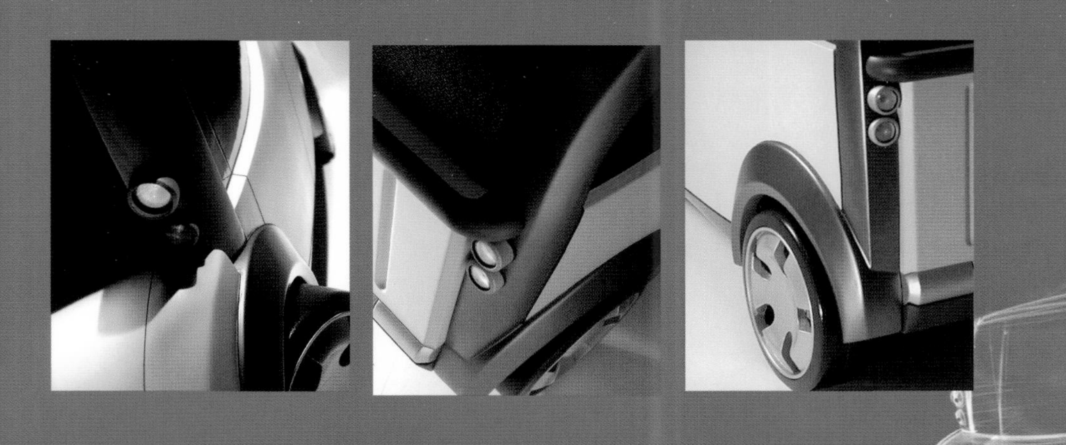

--- studentenfahrzeug auf basis eines sharing-konzeptes (d'n's) -- das fahrzeug ermöglicht hohe flexibilität durch variablen innenraum -- somit ergeben sich je nach sitzkonfiguration drei nutzungsarten: - transporter mit ca 5,5 m^3 ladevolumen; - personentransport: max 8 sitzplätze; - freizeit: bis 5 sitze + laderaum + max. 4 schlafmöglichkeiten (sitzbezüge werden zu hängematten) -- hybridantrieb, radnabenmotoren; offenes fahren durch grosszügige faltdächer; ladefläche durch heckklappe erweiterbar ---

--- a vehicle for students based on a sharing concept (d'n's) -- the vehicle is highly flexible because of its variable interior -- there are three different ways of using the vehicle, depending on the seat configuration: - transport-approx. 5.5m^3 loading capacity; - transportation of passengers: a maximum of 8 seats; - leisure use: 5 seats + loading capacity + a maximum of 4 places to sleep (seat covers used as hammocks) -- hybrid propulsion system, gear hub engine, large folding top gives the vehicle a convertible-like feeling, rear flap makes loading area extendable ---

» e.c(4) «

» roland keplinger, peter körbler «

foto: wallpaper 08/09 2000

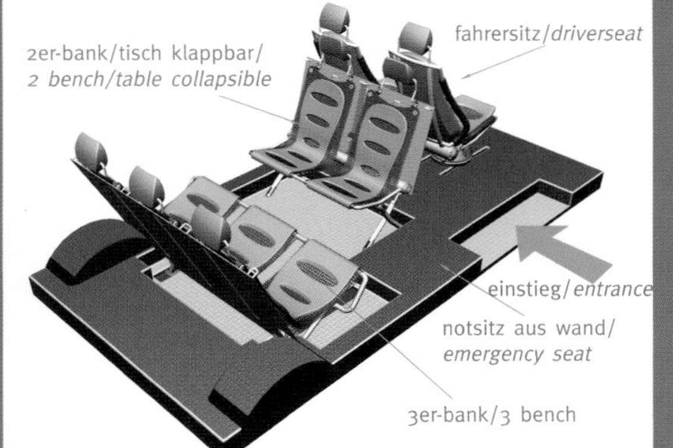

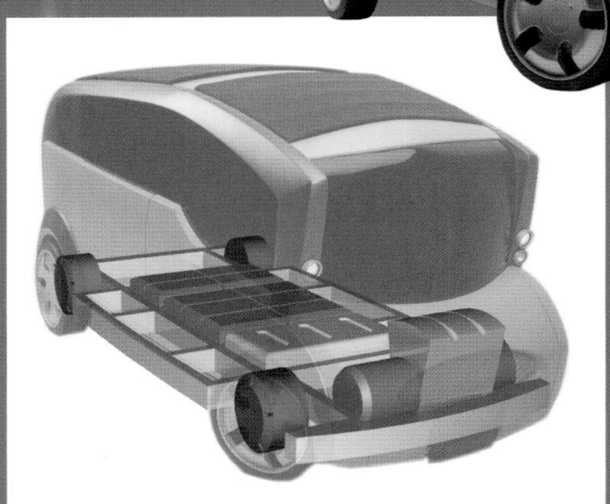

höhenverstellbare kopfstütze/
height adjustable headrest

neopren ummanteltes rohr/
neoprene covered tube

knopf zum verstellen der neigung/
inclining adjust button

drehbar gelagert/*revolving*

» ubique «

» christian koppold, nora langes «

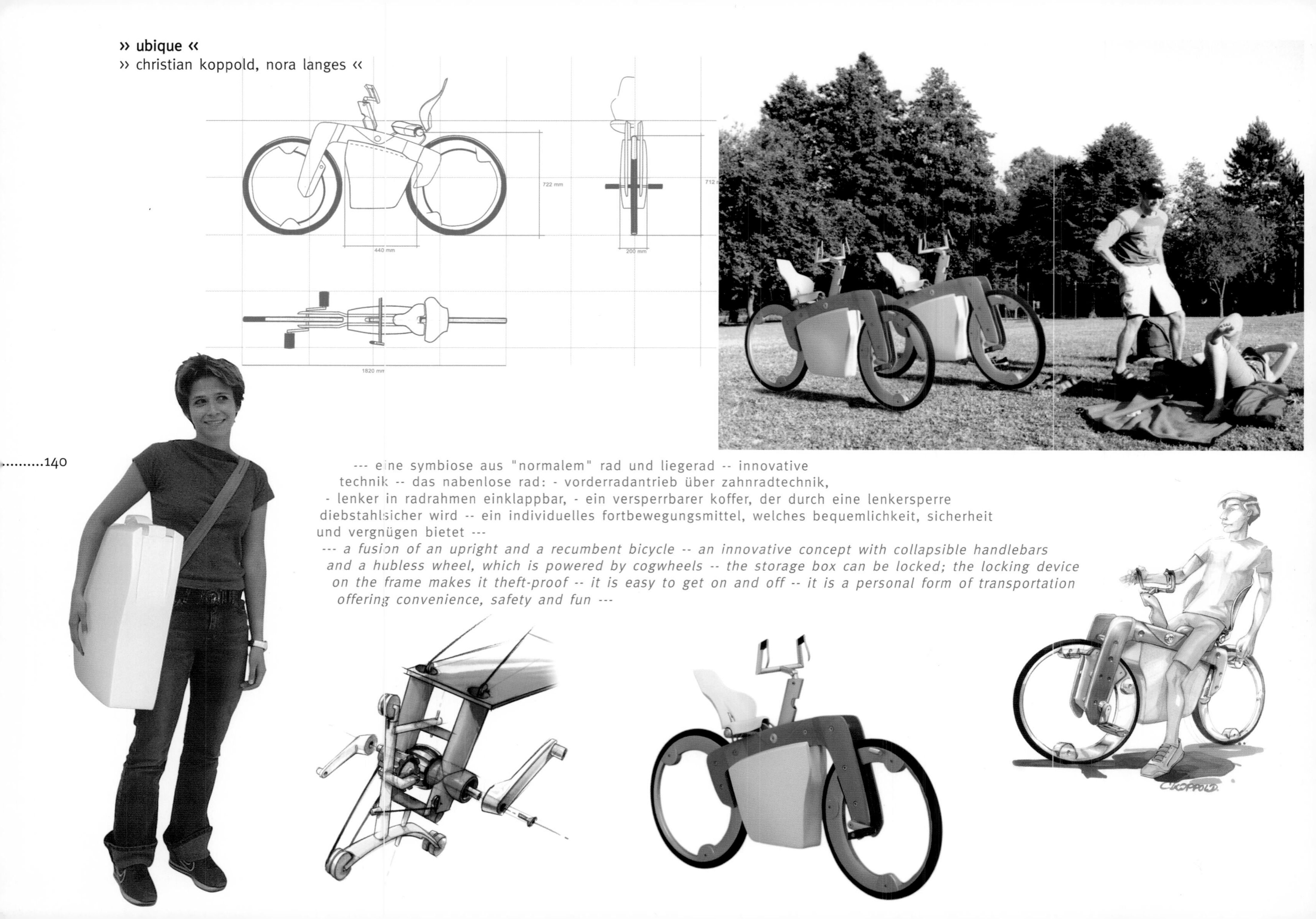

--- eine symbiose aus "normalem" rad und liegerad -- innovative technik -- das nabenlose rad: - vorderradantrieb über zahnradtechnik, - lenker in radrahmen einklappbar, - ein versperrbarer koffer, der durch eine lenkersperre diebstahlsicher wird -- ein individuelles fortbewegungsmittel, welches bequemlichkeit, sicherheit und vergnügen bietet ---

--- a fusion of an upright and a recumbent bicycle -- an innovative concept with collapsible handlebars and a hubless wheel, which is powered by cogwheels -- the storage box can be locked; the locking device on the frame makes it theft-proof -- it is easy to get on and off -- it is a personal form of transportation offering convenience, safety and fun ---

» t.i.e. «

» jan mathis, birgit schaller, rainer trummer «

--- dieses fahrzeug für studenten ist ausgelegt für zwei personen -- besonderes augenmerk liegt auf dem flexiblen transportsystem -- der grosse koffer kann vom fahrzeug abgenommen werden -- dies gibt die möglichkeit, viele kleine dinge auf einmal an einen anderen ort zu bewegen -- maximale geschwindigkeit von 80 km/h -- den eigenen charakter bekommt "t.i.e." durch die kreuzung zwischen nutzfahrzeug und freund auf rädern ---

--- this vehicle is a flexible, motorized transport system for students -- it is constructed as compactly as possible -- one of its special features is a large case that can be removed from the vehicle and taken along -- you can also transport larger objects by attaching them to the vehicle with straps -- this gives you the opportunity to move many smaller objects from one place to another in a single trip -- its maximum speed is 80 km/h -- the character of "t.i.e." is a crossover between utility vehicle and friend on wheels ---

modulares baukastensystem/*modular unit construction system*

kommunikationsbereich/*communication area*

transportbereich/*transport area*

newsroom

chillbereich/*chill area*

» running sushi «

» gabriele bruner, georg hagenauer, elger oberwelz, martin schnitzer «

--- "running sushi" ist der versuch, eine strassenbahn benutzerfreundlicher zu gestalten -- heute übliche massenverkehrsmittel weisen alle eine einheitliche, neutrale gestaltung auf -- dagegen ist "running sushi" in verschiedene bereiche gegliedert, wodurch eine "fahrende landschaft" entsteht -- auf diese weise findet jede benutzergruppe, insbesondere studenten, einen bereich vor, der ihren unterschiedlichen anforderungen bestmöglich gerecht wird ---

--- "running sushi" is an attempt to make a tramway more attractive to the user -- common mass-conveyance today is designed homogenous and neutral all over -- in contradiction to it, "running sushi" is divided into different zones, so that a "moving landscape" evolves -- this is how every user-group, especially students, can find a zone that meets their demands ---

komfortbereich/*comfort area* führerstand/*cab*

2001 (22)

projekt: » diplomarbeiten"
- mit div. kooperationspartnern
8. semester/sommer 2001

betreuer: gerhard h e u f l e r - gerald k i s k a - professoren für design
georg w a g n e r - professor für engineering
werner k l e i s s n e r - walter l a c h - modellbau

thema: die themenbereiche des dritten diplomjahrganges sind durch vielfalt gekennzeichnet. allen gemeinsam ist aber der wunsch nach benutzerorientierten problemlösungen: ob in form eines navigationsgerätes für blinde und sehbehinderte, ob als mitwachsender mountainbike-rahmen für kinder oder als modulare erdbewegungsmaschinen für unterschiedlichste ansprüche. die bandbreite reicht vom intelligenten redesign bis zum innovativen experiment.

project: » theses «
- with different collaborators
8th semester/spring 2001

advisors: gerhard h e u f l e r - gerald k i s k a - professors of design
georg w a g n e r - professor of engineering
werner k l e i s s n e r + walter l a c h - model building

subject: the subjects of this year's theses are marked by diversity. however, a common denominator is the wish for user-oriented problem solutions, whether in the form of a navigational device for blind or vision-challenged people, or a mountain bike frame that grows along with the child or modular earthmoving machines with very different requirements. the range goes from intelligent re-designing to the innovative experiment.

sonic cocoon

ena

chiller

modock

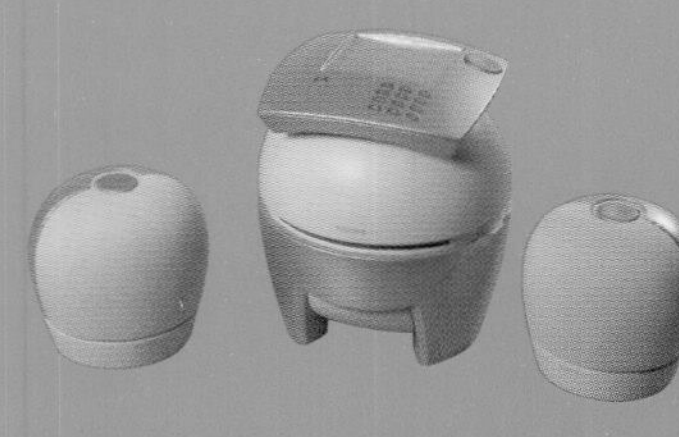
mios

diplom-arbeit

sign

projektarbeiten des fh-studienganges industrial design-graz

projects of the fh-degree course industrial design-graz

2001
(22)

mojito

ted the bike

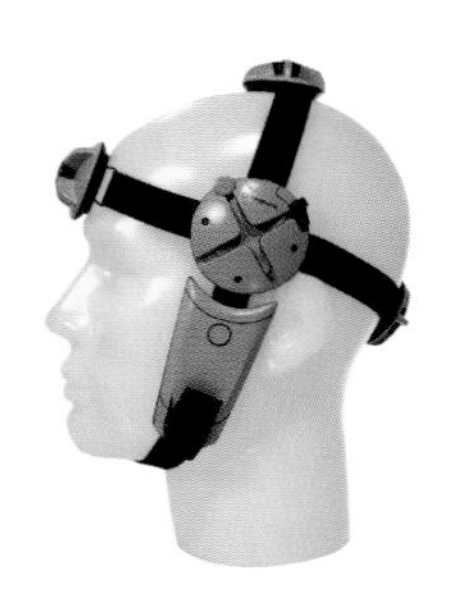

capture

mcr

microcopter

en 2001

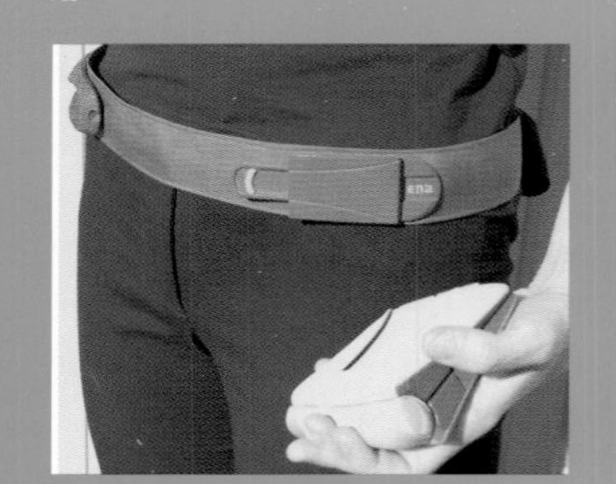

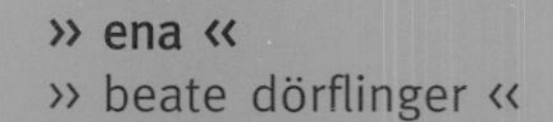

» ena «

» beate dörflinger «

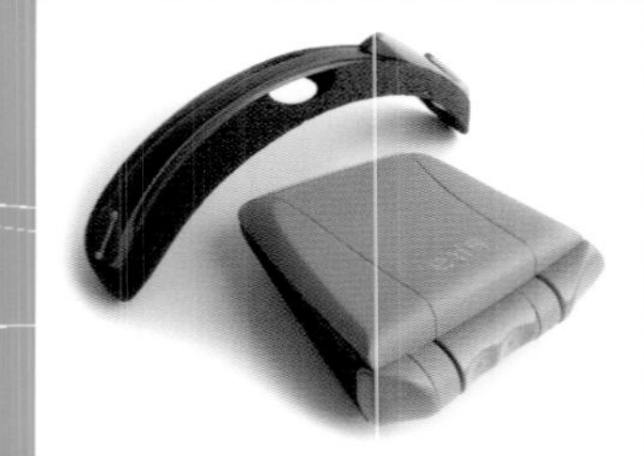

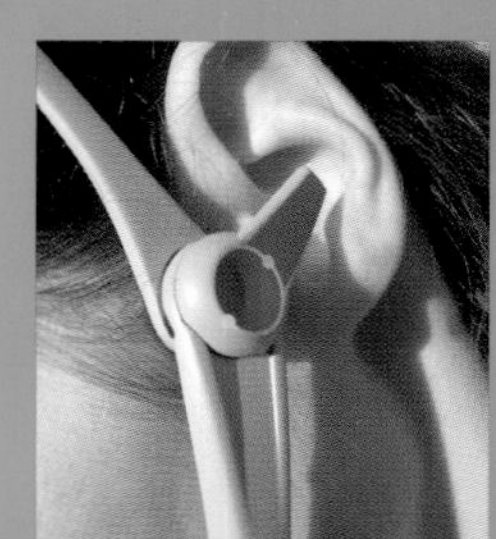

---"ena" = navigator für blinde/sehbehinderte menschen im stadtbereich -- blinde menschen nehmen den raum akustisch wahr, interpretieren ihre umwelt auf grund von geräuschen und klängen -- zur orientierung in der stadt ist jedoch mehr information notwendig -- der navigator kann dem blinden/sehbehinderten menschen nicht den blindenstock ersetzen, aber ihn/sie an jedes gewünschte ziel führen --- ›2.platz mobilitätspreis 2001 zukunftsideen und 3.preis citrix futureaward 2001‹

--- "ena" navigator for blind/visually-impaired people in urban area -- in the city blind people experience and interpret their environment mostly through sounds and noises, but these orientation aids do not tell you where you are -- for orientation in the city more information is necessary -- the navigator cannot replace the cane, but the blind/visually-impaired person can be easily guided to the desired destination ---

›2nd place mobility award 2001 ideas for the future and 3rd place citrix futureaward 2001‹

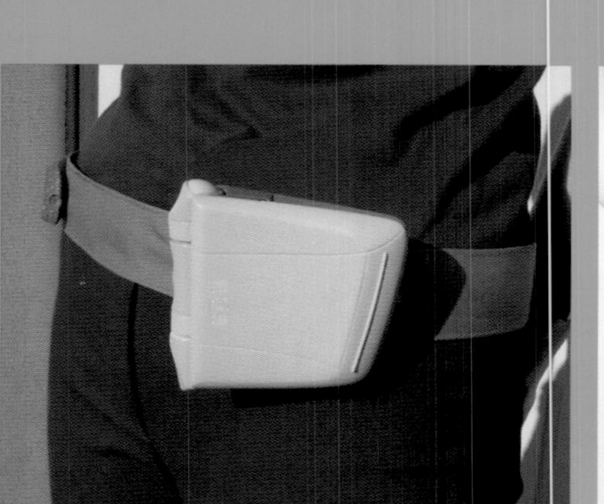

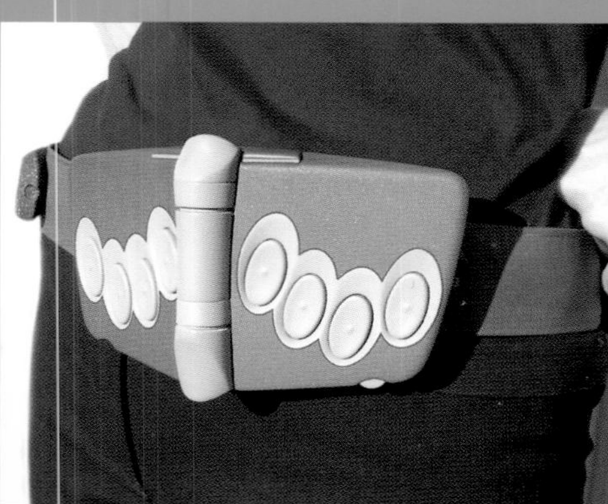

max. rahmengrösse
max. frame size

» ted the bike «
» joe lintl «

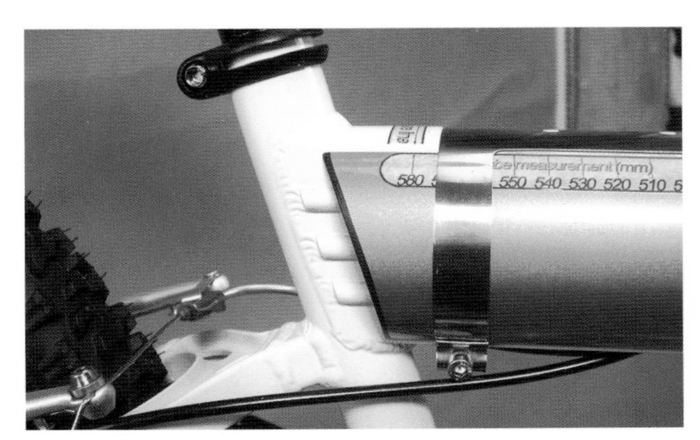

--- "ted the bike", ein mitwachsendes rahmenkonzept für kinder im alter von 8 – 14 jahren --
features: - 1. mitwachsender rahmen; oberrohrlänge 480 bis 580mm; - 2. variable tretlagerhöhe 260 bis 290mm (bringt mehr Fahrsicherheit); - 3. verstellbare lenkervorbau-kombination; - 4. freiliegende kettenstrebe; - 5. durchgehende, innenverlegte brems- und schaltzüge; - 6. stabiler, gleichzeitig leichter aluminiumrahmen ---

--- "ted the bike", a frame that grows with the child, for kids from 8 to 14 years --
features: - 1 frame that grows with the child; - 2 variable pedal bearing height; - 3 adjustable handlebar-stem combination; - 4 exposed chain stay; - 5 continuous, intern routed brake and gear change cables; - 6 stable but light aluminum frame ---

A-A B-B C-C D-D

min. rahmengrösse
min. frame size

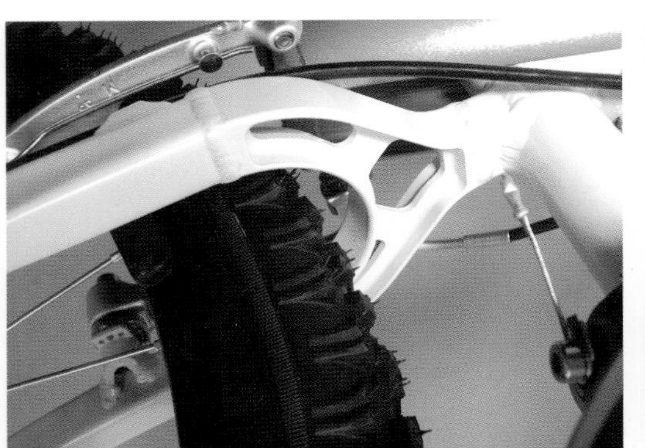

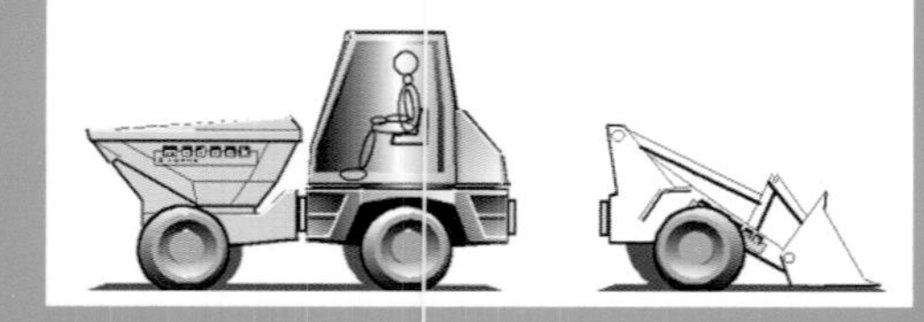

» modock «
» gerald krenn «

--- "modock" ist eine kompakte, modular aufgebaute erdbewegungs-baumaschine mit rahmenknicklenkung und einem neuartigen schnellkupplungssystem, das es erlaubt, verschiedene arbeitsgeräte samt rahmen und rädern direkt am knickgelenk auszuwechseln ---

--- "modock" is a compact, modular earth-moving construction vehicle with frame-center pivot steering and an innovative quick-coupling system that allows a change of different tools including frame and wheels directly at the break joint -- the driver can stay in the cab while changing the equipment ---

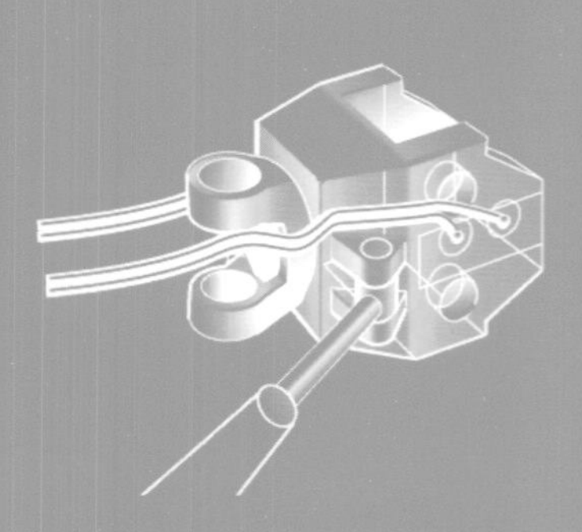

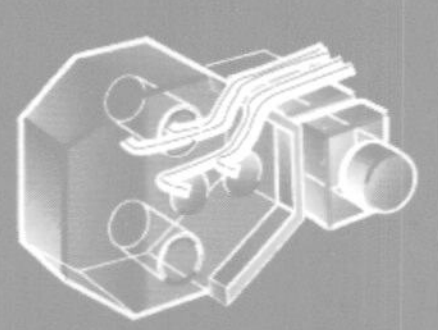

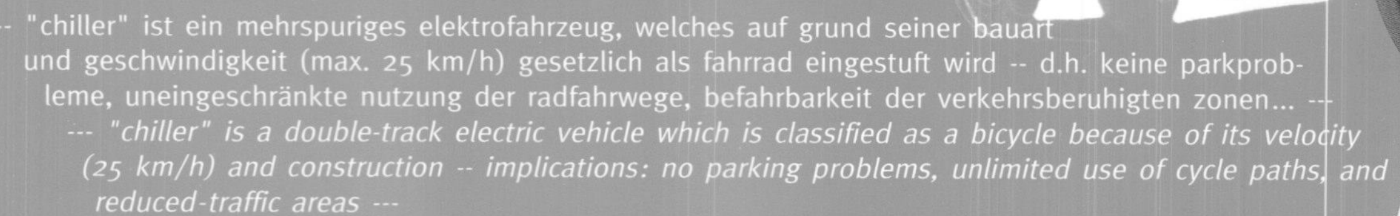

--- "chiller" ist ein mehrspuriges elektrofahrzeug, welches auf grund seiner bauart und geschwindigkeit (max. 25 km/h) gesetzlich als fahrrad eingestuft wird -- d.h. keine parkprobleme, uneingeschränkte nutzung der radfahrwege, befahrbarkeit der verkehrsberuhigten zonen... ---

--- "chiller" is a double-track electric vehicle which is classified as a bicycle because of its velocity (25 km/h) and construction -- implications: no parking problems, unlimited use of cycle paths, and reduced-traffic areas ---

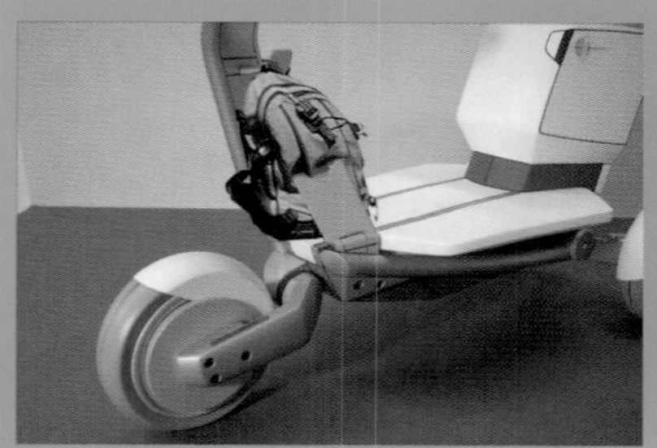

» chiller «
» albert ebenbichler «

» capture - your freedom «
» daniel dockner «

--- diese helmkamera mit abnehmbarem kameramodul und schutzhelm soll den ansprüchen von sportlern gerecht werden -- eigenschaften: - wasserdicht und stossfest; - abnehmbare miniatur-kamera, die am körper und auf ausrüstungsgegenständen montiert werden kann; - speichermedium: kleine securedisc (sd); - handgelenksbildschirm mit flexiblem kunststoffdisplay (polyled-technologie); - vielseitig einsetzbarer schutzhelm vom mountainbiken über klettern bis canyoning... ---

--- this head-mounted camera with detachable camera module and detachable helmet meets the needs of the target group: downhill mountainbiking, climbing, canyoning... -- features: - water- and shock-proof; - detachable mini-camera; - data storage on small secure discs (sd); - wrist-display with a flexible plastic display; - helmet for various applications as mountainbiking and canyoning... ---

» civic suppport urban nomads «
» karin krumphals «

--- reiserucksack, konzipiert für die verwendung im urbanen bereich -- schutz des inhaltes durch faserumwickeltes reflektierendes skelett, sitzmöglichkeit, abdeckung der gurte für flugverkehr, nachziehmöglichkeit, vergrösserbares volumen, versperr- und anhängefunktion, strukturierter innenraum ---

--- travelers' backpack conceived for urban use -- protection of the contents by fibre-reinforced frame, built-in seat, cover for the straps on the back -- the backpack can be pulled like a trolley, extendable, lockable, compartmentation of interior ---

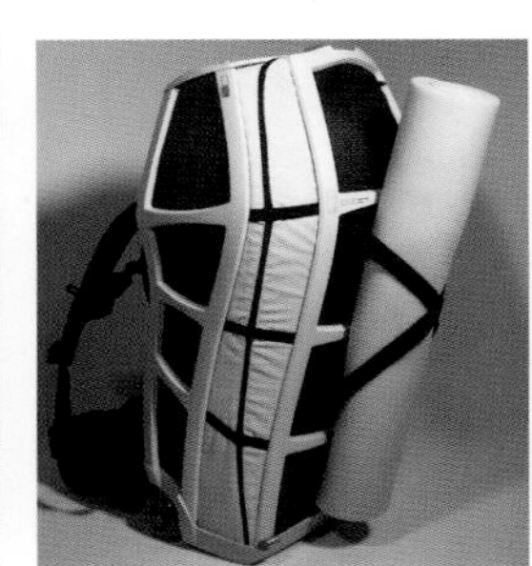

» sonic cocoon «
» peter kalsberger «

--- ein stuhlentwurf, der die möglichkeit bietet, sich mit hilfe von musik zu entspannen -- wie ein kokon schliessen sich die flügeltüren des stuhls um den sitzenden körper, um dadurch einen optimalen klangraum zu erzeugen ---

--- a chair which offers the possibility to relax by means of music -- the folding doors enclose the sitting listener like a cocoon and create the optimal sonic space ---

--- "mios" ist eine modulare audioanlage für zuhause, welche sich in die moderne wohnlandschaft integriert -- das system besteht aus einer basisstation mit digitalem speicher, einer fernbedienung und in mehreren räumen drahtlos installierten lautsprechern -- "mios" kann zu einem "surround sound"-system oder heimkino erweitert werden und ist flexibel ---

"mios" enables the user to freely use a digital (media) data storage pool -- files can be accessed via remote control with a digital display in every room of the home and data is transmitted to the speakers -- all components of the modular system interact wirelessly --- "mios" is designed to blend in with modern interior ---

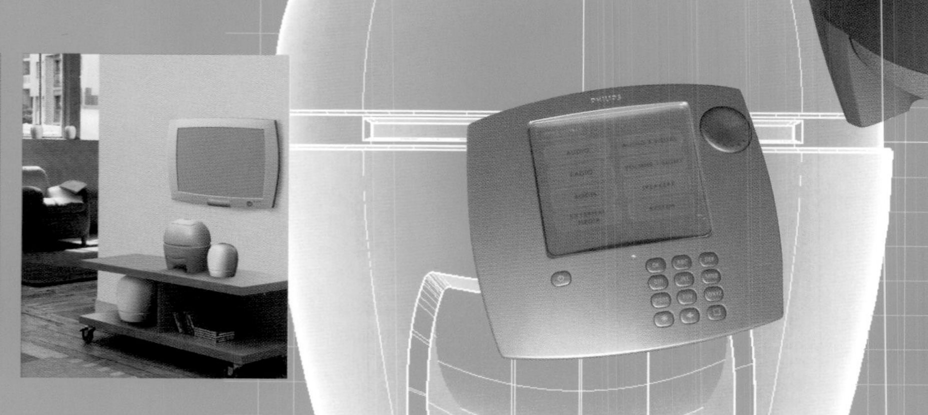

» mios «
» gertraud körner «

» mojito «
» mark ischepp «

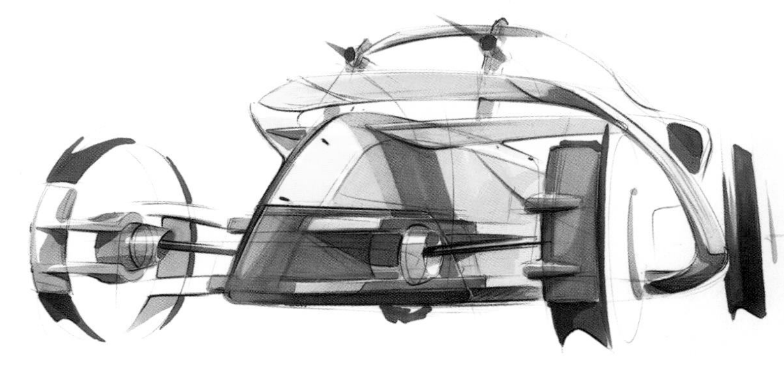

--- formale und technische einheit von interieur und exterieur -- an allen vier rädern baugleiches radaufhängungsmodul -- verminderung der bauteilzahl -- einsatz von unüblichen materialien (z.b.: achsschenkel aus faserverstärktem kunststoff) -- wechselbare stromquelle zum betrieb des elektroantriebs -- annäherung an ein motorradähnliches fahrgefühl ---

--- use of the same formal language for interior and exterior -- multiple use of one suspension component group -- reduction of components -- use of unusual materials like fibre-reinforced plastic for a-arms, etc. -- exchangeable power source to produce energy for the electric engine -- safety level of a car, with open-air feeling of a motorcycle ---

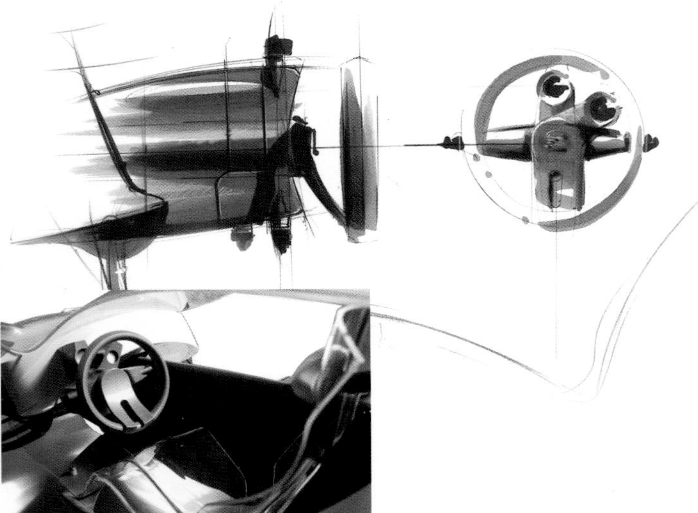

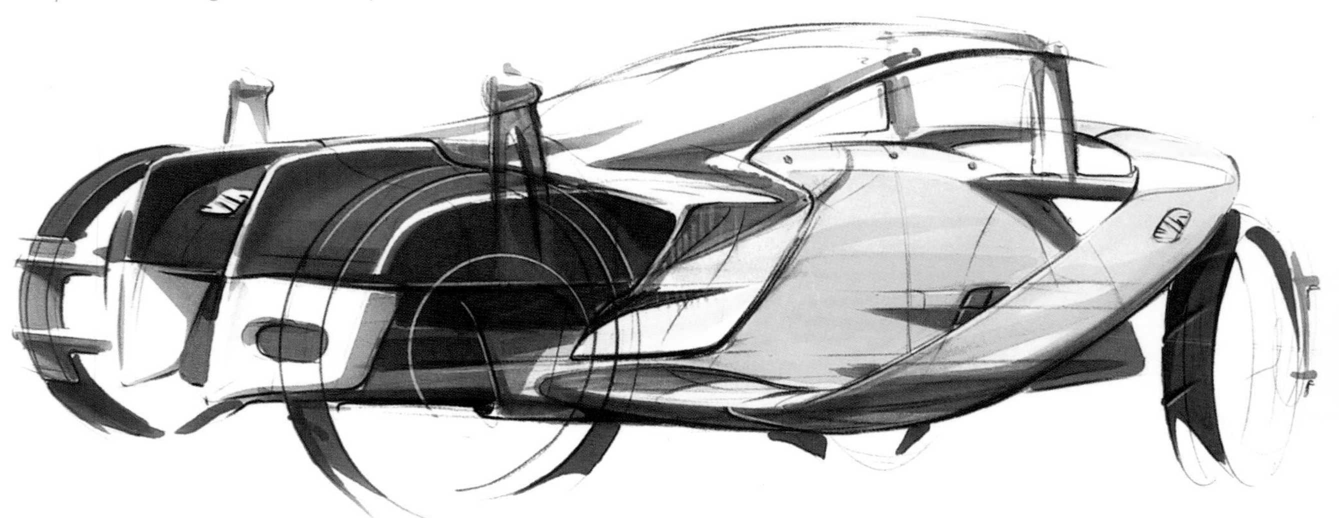

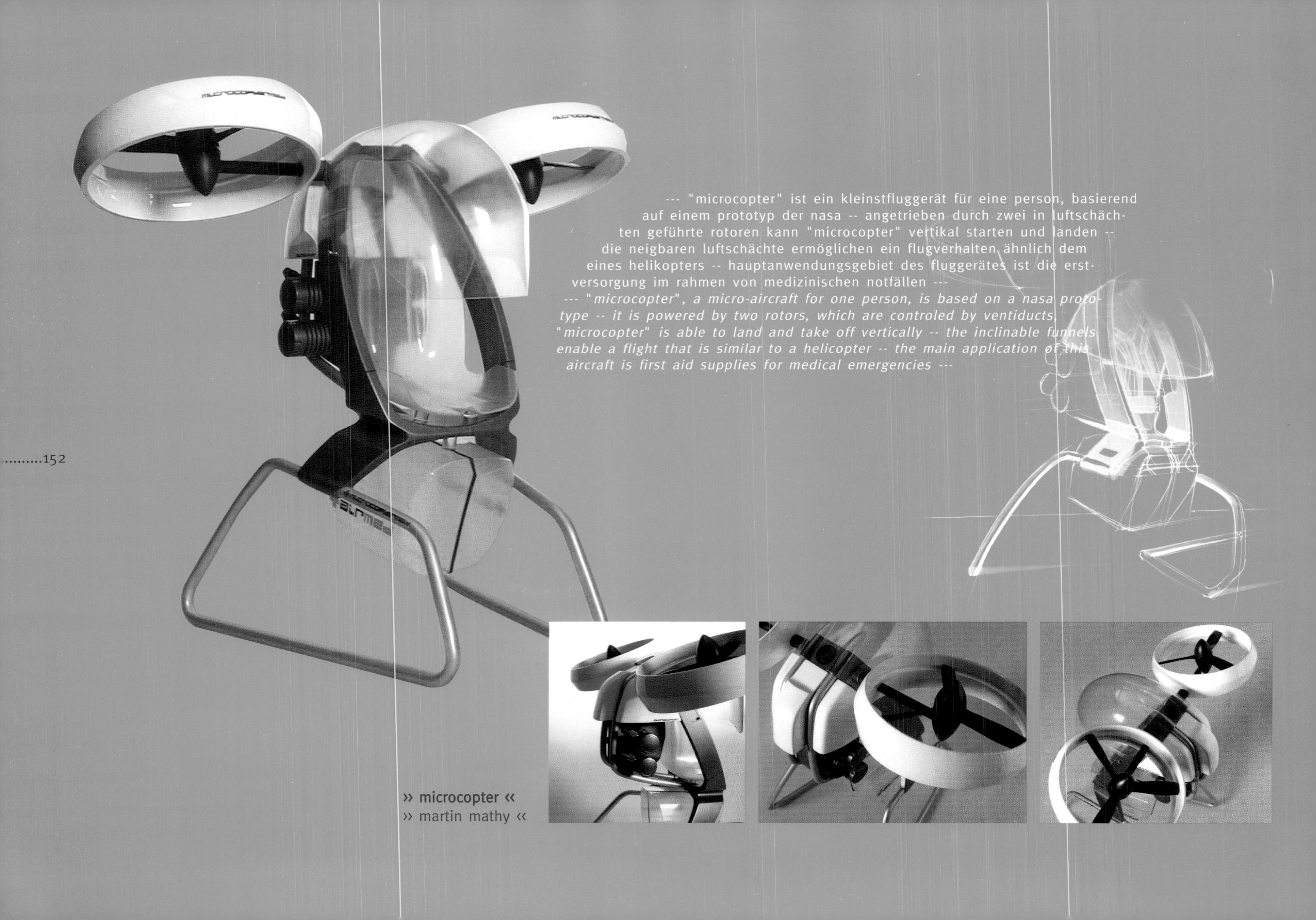

--- "microcopter" ist ein kleinstfluggerät für eine person, basierend auf einem prototyp der nasa -- angetrieben durch zwei in luftschächten geführte rotoren kann "microcopter" vertikal starten und landen -- die neigbaren luftschächte ermöglichen ein flugverhalten ähnlich dem eines helikopters -- hauptanwendungsgebiet des fluggerätes ist die erstversorgung im rahmen von medizinischen notfällen ---

--- *"microcopter", a micro-aircraft for one person, is based on a nasa prototype -- it is powered by two rotors, which are controled by ventiducts, "microcopter" is able to land and take off vertically -- the inclinable funnels enable a flight that is similar to a helicopter -- the main application of this aircraft is first aid supplies for medical emergencies ---*

» microcopter «
» martin mathy «

» mcr - media and communication robot «

» gerfried gaulhofer «

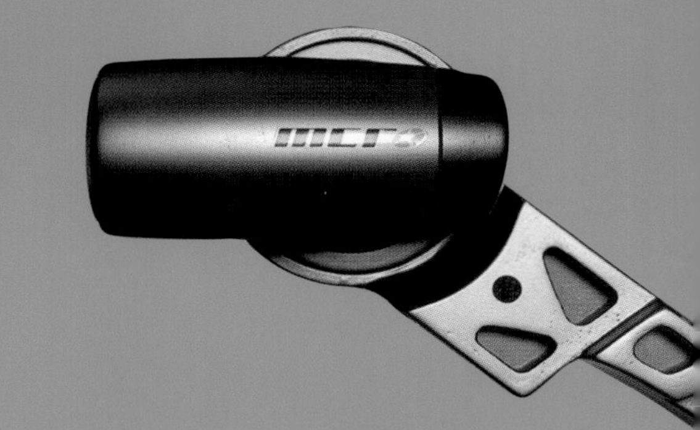

--- der "mcr" ist ein medien- und kommunikationsroboter für den privaten lebensraum -- schnittstellen: - projektor - kamera - mikrofon und - lautsprecher -- die beiden grossen, getrennt angetriebenen räder erlauben es, jede beliebige bewegung durchzuführen, wie z.b. das drehen um die eigene achse -- man geniesst so an beliebigen orten im wohnraum den vollen komfort von internet -- darüber hinaus kann der besitzer eines "mcr" seine wohnung auch aus der ferne beobachten -- im inaktiven (offline) zustand umschliessen die beiden als halbschalen ausgebildeten antriebsräder den roboter -- dieser zustand soll verhindern, dass der user das gefühl hat beobachtet zu werden ---

--- the media and communication robot "mcr" is designed for the living area – interfaces: - projector - camera - microphone and - loudspeakers -- the two separately powered wheels enable the "mcr" to turn around its own axle, which gives its owner the possibility of enjoying the internet at any place of his/her home -- he/she can also watch his/her living space from the distance -- when off-line the computer is enclosed by the two bowl-shaped wheels in order to give the owner some privacy ---

--- generelles fh-ziel: » anwendungsnahe, praxisorientierte ausbildung « ---

--- general fh-goal: » career-oriented and highly practical training « ---

--- unser motto: » ... keine professionellen lehrer, sondern lehrende profis « ---

--- our motto: » ... no professional teachers but teaching professionals « ---

gerhard heufler

fh-prof. dipl. ing. | architekturstudium an der tu graz | 1970-71 produkt-designer bei siemens in münchen | 1971-79 assistent an der tu graz | seit 1975 freiberuflicher industrie-designer in graz (elektronische geräte, schienenfahrzeuge, medizintechnik, umwelttechnik u.ä.) | 1979-95 lehrbeauftragter am mozarteum, salzburg | seit 1995 professor am fh-studiengang industrial design in graz (gründungsmitglied und studiengangleiter)
gerhard.heufler@fh-joanneum.at

fh-prof. dipl. ing. (equivalent to m.s. in engineering) | study of architecture at the tu graz (technical university) | 1970-71 product designer for siemens in munich | 1971-79 assistantship at the tu graz | since 1975 freelance industrial designer in graz (electronic devices, rail vehicles, medical technology, environmental technology, etc.) | 1979-95 lecturer at the mozarteum, salzburg | since 1995 professor at the fh industrial design course of study in graz (founding member and department head)
gerhard.heufler@fh-joanneum.at

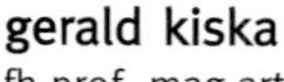

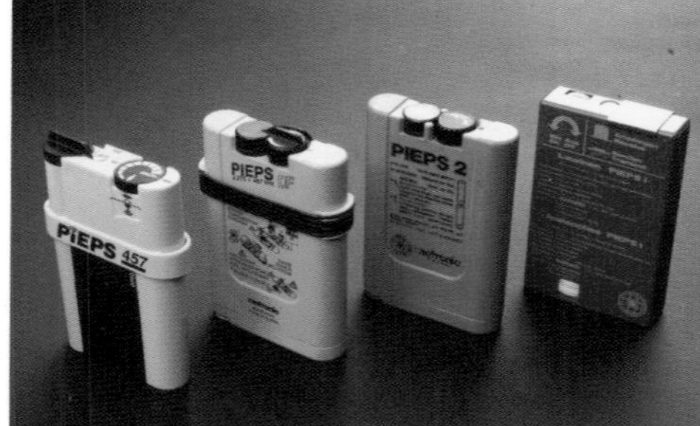

motronic "lawinen-pieps"
notsender und ortungsgerät für lawinenverschüttete
emergency transmitter and detector for avalanche rescue
österr. staatspreis für design 1985
state prize for design 1985

mosdorfer "kabfa"
kabelhalterung für fassaden
cable grips for building facades
österr. staatspreis für design 1985
state prize for design 1985

komptech "topturn 3000"
kompostwendemaschine
compost turning machine
österr. staatspreis für design 1992
state prize for design 1992

siemens "ct 233"
chipkartenterminal
chip card terminal

gerald kiska

fh-prof. mag.art. | designstudium an der hochschule für gestaltung in linz | 1984-85 interform-design, wolfsburg | 1985-87 form orange, götzis | 1987-90 porsche design, zell a. see | seit 1990 inhaber des design büros "kiska creative industries", salzburg (fahrzeuge, konsumgüter, investitionsgüter, grafik design, engineering) | 1994-95 gastprofessor an der hochschule für gestaltung, offenbach | seit 1995 professor am fh-studiengang industrial design in graz (gründungsmitglied)
office@kiska.at

fh-prof. m.a. | study of design at the hochschule für gestaltung in linz | 1984-85 interform-design, wolfsburg | 1985-87 form orange, götzis | 1987-90 porsche design, zell am see | since 1990 owner of the design company "kiska creative industries", salzburg (vehicles, consumer goods, industrial goods, graphics design, engineering) | 1994-95 guest professor at the hochschule für gestaltung, offenbach/germany | since 1995 professor at the fh industrial design course of study in graz (founding member)
office@kiska.at

ktm "duke"
das erste strassenmotorrad von ktm in der zweiten auflage 1999
the first road motorcycle for ktm in its second edition 1999

mke "hydrant"
modularer feuerlöschhydrant 1993
modular fire-extinguishing hydrant 1993

think dig "orderman"
elektronischer "funkbestellblock" für die gastronomie 2001
electronic "radio orderbook" for the catering trade 2001

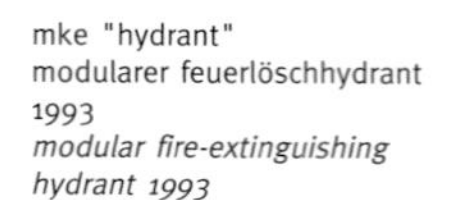

öbb "city shuttle"
nahverkehrswaggon in doppelstockausführung 1995
commuter train in double-decker version 1995

"omni"
dulares blutanalysegerät
dular blood analyser
dikat "design ausgewählt" 1995
sign selected" award 1995

jenbacher energiesysteme "dhe 675"
schienenfräsmaschine
rail milling machine
österr. staatspreis für design 1995
state prize for design 1995

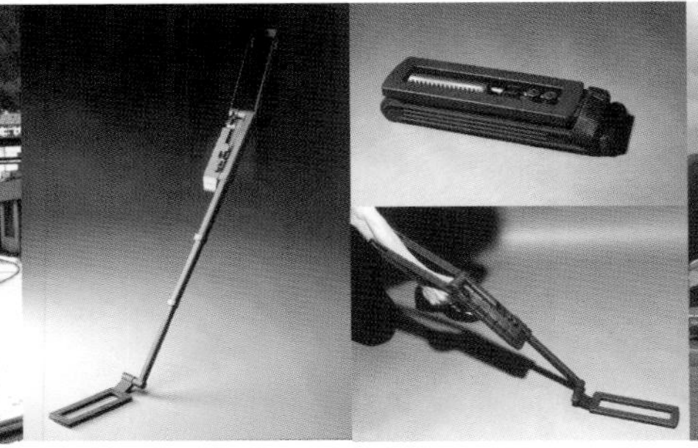

schiebel "mimid"
miniatur minensuchgerät
miniature mine detector
staatspreis für design 1997/98
industrial design excellence award 1998 usa: silver
honorable mention award i.d. annual review 1998 usa
"class a - österreichische produktkultur heute" 1998
honorable mention award design-biennale bio 1998
time magazin usa: best of 1998 design (10.)
design preis schweiz 1999: anerkennungspreis
honorable mention award
"international design yearbook 1999/2000" gb

integral verkehrstechnik "integral"
nahverkehrszugsystem
shuttle train system
prädikat "design ausgewählt" 2001
"design selected award" 2001

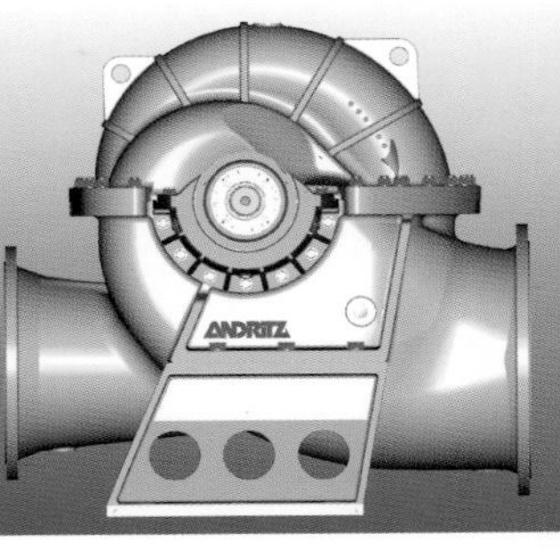

andritz "fp 60-500"
mischpumpe
fan pump

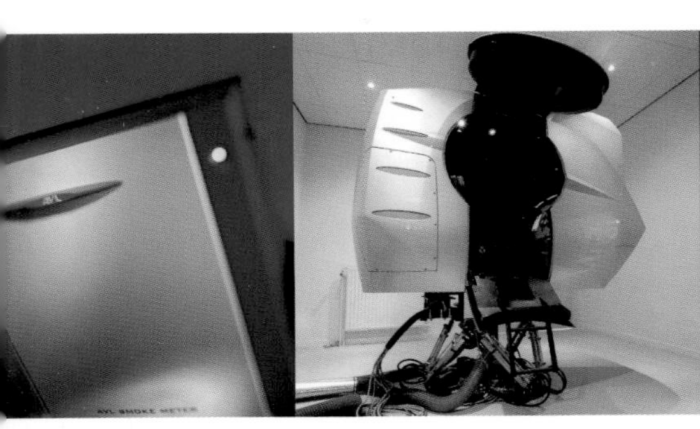

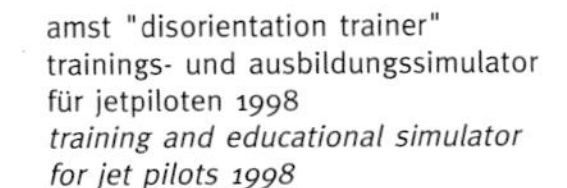

"smoke meter"
gasmessgerät zur
torenentwicklung 2000
haust emission gauge for
otor development 2000

amst "disorientation trainer"
trainings- und ausbildungssimulator
für jetpiloten 1998
training and educational simulator
for jet pilots 1998

ktm "unit"
motorradstudie 1996
motorcycle design study 1996

cadence "arca"
elektrostatischer
hybrid-lautsprecher 1994
electrostatic
hybrid loudspeaker 1994

stork "laser-engraver 5102"
lasergravurmaschine zur druckmatrizenerzeugung 2001
laser engraving machine for the production of print matrix 2001

multiform "concept 2"
multifunktionaler konzepthalter
1997
multifunctional concept holder
1997

fh joanneum graz - 1920 errichtet als waggonfabrik - 1997 als fachhochschule adaptiert (architekten eisenköck & peyker)

fh joanneum graz - built as a railroad car factory in 1920 - adapted for fachhochschule in 1997 (architects eisenköck & peyker)